THE
CAPACITOR HANDBOOK

Second Edition

Cletus J. Kaiser

*Published
by*

CJ Publishing
398 Wintergrape Ln.
Rogersville, MO 65742

©Copyright 1990, 1995 by Cletus J. Kaiser, Olathe, KS 66061.

SECOND EDITION FIRST PRINTING—1995
SECOND PRINTING—1997

All rights reserved. No part of this book shall be reproduced, stored in a retrieval system, or transmitted by any means, electronic, mechanical, photocopying, recording, or otherwise, without written permission from the publisher. No patent liability is assumed with the respect of the information contained herein. While every precaution has been taken in the preparation of this book, the publisher and author assume no responsibility for errors or omissions. Neither is any liability assumed for damages resulting from the use of the information contained herein.

ISBN: 0-9628525-3-8
Library of Congress Catalog Card Number: 94-094595

Printed in the United States of America.

Table of Contents

Acknowledgements v
Preface vi

Chapter 1   Fundamentals For All Capacitors   1

    Application Information 21

Chapter 2   Ceramic Capacitors   29

    Application Information 38

Chapter 3   Plastic Film Capacitors   47

    Application Information 53
        Plastic Film Capacitors 53
        Metallized Film Capacitors 54

Chapter 4   Aluminum Electrolytic Capacitors   57

    Production Technology 61
        The Anode (Positive Plate) 61
        The Electrolyte 63
        The Spacer 65
        The Cathode 65
    Electro-mechanical Considerations 66
    Application Information 71

Chapter 5   Tantalum Capacitors   79
    Tantalum Foil Style 80
    Wet Tantalum Style 81
    Solid Tantalum Style 81
  Application Information 87
    Tantalum Foil Capacitors 87
    Wet-Electrolyte, Sintered Anode Tantalum Capacitors 89
    Solid Tantalum Capacitors 92

Chapter 6   Glass Capacitors   97
  Application Information 100

Chapter 7   Mica Capacitors   103
  Application Information 105

Glossary   107

Bibliography   117

Appendix A   Capacitor Selection Guidelines   119
  Ceramic 119
  Paper/Plastic Dielectric 120
  Aluminum Electrolytic 122
  Tantalum Electrolytic 123
  Glass 124
  Mica 124
  Trimmer Capacitors 124

Appendix B   Equations and Symbol Definitions   125
  Basic Capacitor Formulas 125
  Metric Prefixes 128
  Symbols 128

Index   129

Acknowledgments

The author is thankful to The Lord.

The author is deeply indebted to his family for their guidance and support.

Through the courtesy of Matthew Pobursky, with his publishing skills, the author gratefully acknowledges his professional help and services in making this book possible.

Appreciation is expressed to the many friends, both in the technical and publishing communities, who made specific suggestions concerning content and organization of this book.

Preface

This second edition was written as an update after the publication of the book titled "The Resistor Handbook."

A long and varied experience in many areas of electronic circuit design has convinced me that capacitors are the most misunderstood and misused electronic component. This book provides practical guidance in the understanding, construction, use, and application of capacitors. Theory, combined with circuit application advice, will help to understand what goes on in each component and in the final design.

All chapters are arranged with the theory of the dielectric type discussed first, followed by circuit application information. With all chapters arranged in the same manner, this will make reading and using this book for reference easier. A practical glossary of terms used in the capacitor industry is included.

The first chapter covers basic information that applies to all types of capacitors. Each following chapter addresses a different capacitor dielectric. This book could have been titled: 'Everything You Wanted To Know About Capacitors, But Were Afraid To Ask ...'

Chapter 1

Fundamentals For All Capacitors

For all practical purposes, consider only the parallel-plate capacitor as illustrated in Fig. 1.1—two conductors or electrodes separated by a dielectric material of uniform thickness. The conductors can be any material that will conduct electricity easily. The dielectric must be a poor conductor—an insulator.

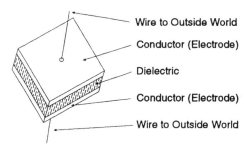

Fig. 1.1   The Parallel-Plate Capacitor

Fig. 1.2 illustrates the symbol for a capacitor used in schematic diagrams of electronic circuits. The symbol resembles a parallel-plate model.

Fig. 1.2   Capacitor Symbol

Fig. 1.3 is a sample circuit that contains all the components normally called "passive," plus a battery. The battery is an "active" component because it can add energy to the circuit. Passive components may store energy momentarily, but they cannot add energy on a continuous basis. The three main passive devices are resistors, capacitors, and inductors.

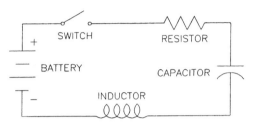

Fig. 1.3   Passive Series Circuit with Battery

A favorite analogy compares the flow of electric current with the flow of water out of a tank as in Fig. 1.4. A capacitor stores energy when it is charged. The water tank would be the capacitor and it would be charged by a pump (a battery) that fills it up. The amount of charge in the capacitor would be analogous to the amount of water in the tank. The height of the water above some reference point would be the voltage to which the battery had pumped up the capacitor, and the area of the tank would be capacitance. A tall, skinny tank might contain the same amount of water as a shallow, flat tank, but the tall, skinny tank would hold it at a higher pressure. Other possibilities are tall, skinny capacitors (high voltage, low capacitance) and shallow, flat capacitors (low voltage, high capacitance).

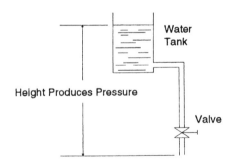

Fig. 1.4   Water Flow Analogy

When the valve of Fig. 1.4 is opened, water runs out. The valve is both a switch and a resistor. If the valve is opened only partially, it causes enough friction so that the water runs slowly from the tank. It is thus like a variable resistor. When resistance is high, the water runs slowly, but if resistance is made small, the water can run more freely. Once the water is running, it can be stopped by closing the valve. The water in the pipe, already in motion, must stop. When closing the valve very quickly, the water must stop flowing very quickly. The energy in the moving water suddenly has no place to go. In some plumbing systems, a distant "chunk" is heard when a valve is closed quickly. The energy in the moving water suddenly has no place to go, so it bangs a pipe against its support somewhere. This is called "water hammer." The moving water has acted like the inductor in the electronic circuit of Fig. 1.5. The battery is the pump, the capacitor is the tank, the resistor and the switch are the valve, and the inductor is the moving water in the pipe.

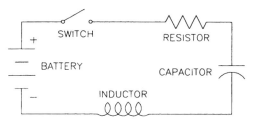

Fig. 1.5 Passive Series Circuit with Battery

Fig. 1.6 illustrates what happens inside a capacitor. When charged by a battery, one electrode of the capacitor will obviously become positively charged and the other one will be correspondingly charged negatively.

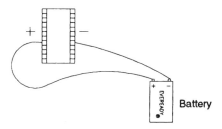

Fig. 1.6 Charged Capacitor

Magnifying the diagram of the capacitor a little bit, Fig. 1.7 illustrates that the presence of electrical charges on the electrodes induces charges in the dielectric. These induced charges determine something called permittivity. Each different dielectric material has its own value of permittivity. Permittivity introduces a more practical and better known value called "K" or dielectric constant. K is the ratio of the permittivity of the dielectric in use to the permittivity of free space—a vacuum. Therefore, all the capacitance values are related to the permittivity of a vacuum.

Fig. 1.7 Charges Inside the Capacitor

In a vacuum, K = 1, while K in every material has some value greater than 1. The higher the K, the more capacitance with all other variables being equal.

Fig. 1.8 is the expression of capacitance. The constant, 8.85 x 10^{-12}, is the permittivity of vacuum.

$$C = (8.85 \times 10^{-12}) K \frac{A}{D}$$

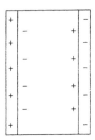

Fig. 1.8 The Capacitance Equation

With this equation, the units must be: capacitance in farads (named for Michael Faraday), the area (A) in square meters and the distance between electrodes (D) in meters. K is simply a ratio and a pure number without dimensions. When units other than farads and meters are used, different constants are used: microfarads and inches for example.

To get an idea of what a farad is, calculate the area which would be necessary in a capacitor built to have one farad, to operate in a vacuum, and to have a spacing between electrodes of one millimeter. First, turn the equation around to solve for the area and then plug in the known values. This calculates to 113 million square meters that would be a field about 6½ miles on a side.

$$C = (8.85 \times 10^{-12}) K \frac{A}{D} \quad or \quad A = \frac{CD}{(8.85 \times 10^{-12}) K}$$

Given: $K = 1$
$C = 1$ farad
$D = 1$ millimeter (or 0.001 meters)

$$A = \frac{1 \times 0.001}{(8.85 \times 10^{-12}) \times 1} = 113{,}000{,}000 \text{ sq. meters}$$

That is why one farad capacitors aren't made very often and when they are, they are never made with a vacuum dielectric and a one millimeter spacing. Industry does make vacuum capacitors, but the market is limited to laboratory standards. All commercial capacitors use some different dielectric material with a higher value of K.

Fig. 1.9, shown on the following page, is a table for dielectric materials that are generally used today. Note a tendency toward the higher values of K for reasons that are now obvious. (With a K of 10, that one farad capacitor area can be reduced to a mere 11.3 million square meters!) The wide range of values for barium titanate, which is the basis for most ceramic capacitors, is an unfortunate fact of nature which will be discussed more completely later. A typical question is why industry makes commercial capacitors with any of the materials having low values of K. The answer generally lies with other capacitor characteristics such as stability with respect to temperature, voltage ratings, etc. These will all be explored as we proceed with particular dielectric systems in the following chapters.

Dielectric Constants for Common Insulators at 25°C	
Insulator	K Value
Air or Vacuum	1.0
Paper	2.0 – 6.0
Plastic	2.1 – 6.0
Mineral Oil	2.2 – 2.3
Silicon Oil	2.7 – 2.8
Quartz	3.8 – 4.4
Glass	4.8 – 8.0
Porcelain	5.1 – 5.9
Mica	5.4 – 8.7
Aluminum Oxide	8.4
Tantalum Oxide	26
Barium Titanate	1,000 – 3,000
Ceramic	12 – 400,000

Fig. 1.9   Table of Dielectric Constants

To understand the behavior of capacitors when connected in a circuit, probably the simplest is the RC timing circuit shown in Fig. 1.10. It is called an RC circuit because the combinations of resistance (R) and capacitance (C) determine its operation.

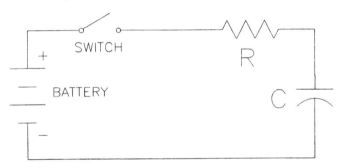

Fig. 1.10   The RC Timing Circuit

When the switch is closed, current from the battery flows through the circuit, charging the capacitor. When the capacitor is completely charged, it is like a closed tank which is completely filled up, and no further current flows. At that time, the voltage across the capacitor would be equal to the supply voltage of the battery. Voltage across the capacitor advances from zero (fully discharged) to the supply voltage along some predetermined path with respect to time. If the resistor is small, current flows easily and the capacitor is charged more quickly. If the resistor is very large, the charging process follows a different path and will take longer to complete.

The behavior of voltage versus time is also influenced by the size of the capacitor. If the capacitor's capacitance is very large, it will require more total energy to fill (the tank is larger in diameter), and current flowing through the resistor will require a longer time to charge it. Fig. 1.11 illustrates three charging curves, each approaching the same end point but along different paths.

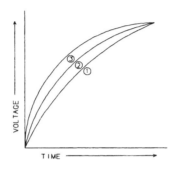

Fig. 1.11   Voltage Across Capacitor in RC Circuit

By adjusting the value of resistance in R and the capacitance in C, formation of curves 1,2,3 and many others can be obtained. A typical application of this circuit might be to leave the lights on in your car and have them go off automatically after you are inside the house. The voltage across the capacitor can be used to operate a switch when it reaches some predetermined value. If other considerations in this circuit required that the switch be operated on a decreasing voltage rather than an increasing voltage, the voltage which appears across the resistor in the circuit can be used, as shown in Fig. 1.12 on the following page.

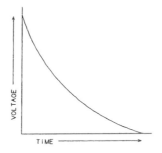

Fig. 1.12 Voltage Across Resistor in RC Circuit

The instant the switch is closed, all the voltage of the battery would appear across the resistor and none across the capacitor. The voltage across the resistor would decrease with time just as the voltage across the capacitor increases with time.

The timing circuit is a good example of a DC application. Note that the capacitor blocks flow of DC once it is charged. Current would flow once more if another switch was connected to discharge the capacitor, as in Fig. 1.13. If switch 1 is opened and then switch 2 is closed, the stored energy in the capacitor would flow as current through the resistor until the voltage across the capacitor reached zero. The capacitor can thus be compared to a storage battery, although the principles of operation are entirely different.

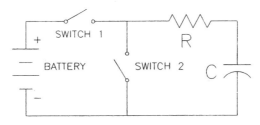

Fig. 1.13 Discharge Through Switch 2

The storage capability of the capacitor is used to good effect in filters. A typical DC power supply offers a good case for an example. Basic DC power supplies provide an output (that is, the voltage across a load, shown in Fig. 1.14 as a resistor) which is fluctuating.

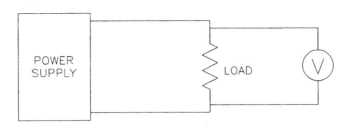

Fig. 1.14 Measuring Voltage From A Power Supply

Fig. 1.15 shows a situation where the voltage drops completely to zero. What is really wanted is a straight line across this graph representing a steady DC voltage.

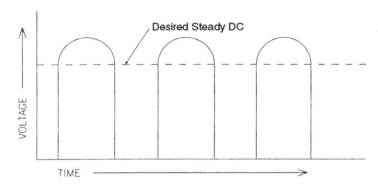

Fig. 1.15 Fluctuating DC Voltage From A Power Supply

To approach the desired straight line, add a capacitor to the circuit to smooth these fluctuations as shown in Fig. 1.16.

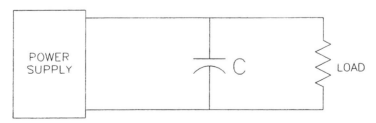

Fig. 1.16 Filter Capacitor Added

With the voltage at zero and the capacitor discharged, turn the supply on. As the voltage begins to rise, some current will flow to charge the capacitor while the rest passes through the resistor. Some time before the capacitor is completely charged, the voltage from the supply will begin to decline. As soon as the supply voltage is below the capacitor voltage, the capacitor will begin to discharge, and current will flow from the capacitor, tending to maintain the voltage across the resistor. If the value of capacitance is chosen correctly, the capacitor cannot be totally discharged during the time available, and the capacitor will be charged once more as the supply voltage exceeds the capacitor voltage.

The result of a simple filter of this sort will not produce the desired steady DC voltage (a perfectly straight line on the graph), but it will produce a waveform something like that seen in Fig. 1.17.

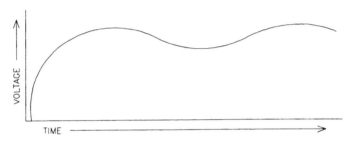

Fig. 1.17   Filtered DC

The condition can be improved further by adding a series resistor and another capacitor as shown in Fig. 1.18.

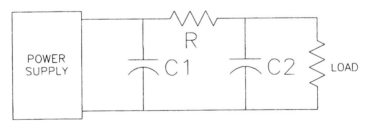

Fig. 1.18   Improved Filtering

An even better result can be obtained if an inductor is used instead of the series resistor as shown in the circuit of Fig. 1.19. (Remember the water in the pipe which wanted to keep running?)

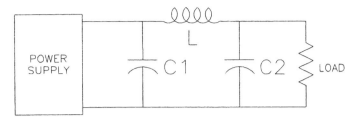

Fig. 1.19   Even Better Filtering

Alternating current must also be considered. Here, the voltage goes from zero to some maximum value, back down to zero, and then in the negative direction before returning to zero once more. Alternating current frequently does look like that in Fig. 1.20, which is a sine wave.

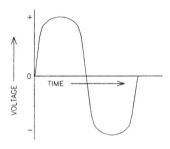

Fig. 1.20   Alternating Current - The Sine Wave

When a capacitor is subjected to alternating current, to the capacitor, it looks just like DC which is flowing in and flowing out again. The capacitor is alternately being charged, discharged, and then recharged in the opposite direction before being discharged again. One important fact to note is that the capacitor can never block the flow of AC but instead permits a steady flow of current. This throws the timing circuit out the window, of course, but it opens up a lot of new possibilities.

Consider how much current flows through the circuit shown in Fig. 1.21. If the generator's sine wave voltage and the resistance, R, don't change, current flow depends upon only capacitance and frequency.

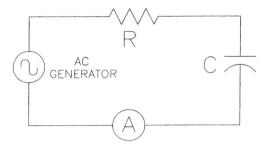

Fig. 1.21 Current Flow In An AC Circuit

If it is a large capacitor, there will be insufficient time for it to be charged more than a small amount before the current direction reverses and it is discharged again. Current flows very easily when the capacitor is near its discharged state, as we noted with the timing circuit. If the capacitor is small, it might approach the completely charged state before the current reverses direction and discharges it. The smaller capacitor would thus offer much more hindrance to the passage of current.

The second factor affecting current flow is the frequency of the alternating current. If, instead of the previous wave form, the wave form is one in which current reversal takes place in half the time (double the frequency) as in Fig. 1.22, the amount of energy which flows into the capacitor before current reversal will be be much less. In effect, the capacitor will stay closer to its discharged state than when the frequency of the waveform which the capacitor offers will be less.

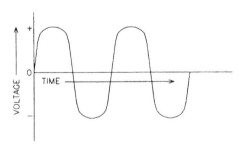

Fig. 1.22 Higher Frequency AC

The capacitor, in an AC circuit, is acting something like a resistor in a DC circuit with the additional dimension of frequency to take into consideration.

The two effects of frequency and capacitance are combined in an expression known as capacitive reactance and is expressed as "X_C." Note that X_C is expressed in ohms (Ω), which is the unit of resistance. Reactance acts *something* like resistance, and uses the same unit in order to combine the two later. The frequency is expressed as the number of alternations (complete cycles) which occur in one second, abbreviated cps for "cycles per second" or Hz for "Hertz." Note that capacitive reactance is *inversely* proportional to both frequency and capacitance. This fits exactly with the earlier explanation concerning the ease of charge and discharge of a capacitor when is was operating near its discharge state.

The formula for determining X_C can be expressed as follows:

$$X_C = \frac{1}{2\pi f C}$$

Where: X_C = capacitive reactance, Ohms
π = 3.14
f = frequency, Hertz (cycles per second)
C = capacitance, Farads

There is a comparable expression for inductance which yields inductive reactance. The unit of inductance is the Henry. It follows that inductance in an AC circuit impedes the flow of the current just as a capacitor does. The difference is that X_L is directly proportional to both frequency and inductance. The larger the inductor and the higher the frequency, the greater is the reactance to current flow: just the opposite of the behavior of a capacitor's X_C.

The formula for determining X_L can be expressed as follows:

$$X_L = 2\pi f L$$

Where: X_L = inductive reactance, Ohms
π = 3.14
f = frequency, Hertz (cycles per second)
L = inductance, Henries

Unfortunately, capacitors include both resistance and inductance, because it is not possible to make practical devices completely lacking in these factors.

Fig. 1.23 is a series circuit with all three passive component properties: resistance, inductance and capacitance, and all capacitors actually look something like this.

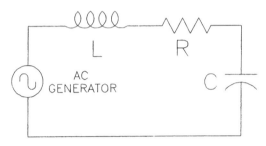

Fig. 1.23 Passive Elements in a Series AC Circuit

Of course, if the capacitor is a good one, the amount of resistance and inductance is very small compared to the amount of capacitance. In an AC circuit, all three components act to decrease the flow of current. The sum effect of all three is termed "impedance." Impedance is expressed in ohms just like resistance, and it would be nice if we could simply add X_L, X_C, and R to get impedance, Z. Unfortunately, it isn't quite that simple, X_C and X_L can be added or subtracted directly, but they must be combined with R by the squaring and square root process indicated here.

$$Z = \sqrt{R^2 + (X_L - X_C)^2}$$

Many capacitor catalogs show graphs of impedance versus frequency. Impedance becomes a very useful consideration at higher frequencies because the capacitive effect disappears at some frequency, dependent on capacitor design. Remember, X_C decreased as frequency was increased. Also remember that X_L increased as frequency increased. So, capacitors are built which have very little inductance and a lot of capacitance, but if the frequency is raised to a high enough value, X_C is eventually overtaken by X_L and the capacitive device now acts like an inductor.

By plotting impedance with logarithmic scales, we get a graph which looks like Fig. 1.24. Where the capacitive reactance is equal to the inductive reactance, this is called the "self-resonant" point. If there were no resistance in the circuit, the impedance would drop to zero at this point.

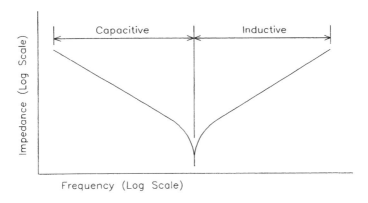

Fig. 1.24   Impedance With Logarithmic Scales

Practical capacitors frequently look more like Fig. 1.25 because they do include resistance. Exactly at the self-resonant point, in fact, they act entirely like a resistor.

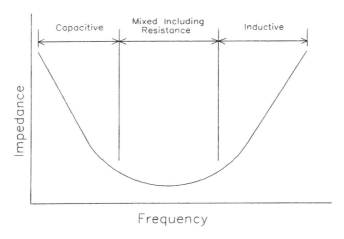

Fig. 1.25   Impedance vs. Frequency

Fundamentals For All Capacitors

Pure resistance does not change with frequency. In practical capacitors, however, the simple series circuit does not exist, but rather there is a fairly complex mixture of resistance, capacitance and inductance. The result is that, as measured at the terminals of the capacitor, we see a resistance which declines with frequency. Because it really is not a pure resistance, it is called ESR or "Equivalent Series Resistance." Manufacturers' capacitor catalogs give graphs of ESR for many capacitors.

Inductance, which arises primarily from the fact that lead wires are attached to the capacitors, is seldom a problem at low frequencies, although as clock rates of computers increase, manufacturers are now designing capacitors to minimize inductance. Resistance, however, is quite often a problem because it limits the power handling capability of the capacitor. An ideal capacitor (or an ideal inductor for that matter) would produce no heat when current passes through it. The heat which is produced in practical devices comes from the resistance which manufacturers are unable to eliminate completely. Because it has this importance, a measure of the resistance is frequently specified. We could use ESR directly, but it has been found much more convenient to use an expression called "DF" (dissipation factor). The expression for DF is the ratio of resistance to capacitive reactance. The higher the resistance, the higher the DF and generally the worse the capacitor. Because in good capacitors the DF is rather small, it is frequently expressed in percent. Most people would rather read 3% than 0.03.

The dissipation factor of a capacitor can then be expressed as:

$$DF = \frac{R}{X_C}$$

By using DF rather than ESR, we can have one factor which represents a measure of capacitor quality applicable to a fairly wide range of capacitance values. If ESR was used, it would be necessary each time to specify the value of capacitance.

You may also see the expression for DF written this way. This is simply the result of substituting the component factors of X_C which we saw earlier.

$$DF = 2\pi f C R$$

Capacitors in electronic circuits are normally not subjected to very large AC currents. Upon occasion, however, the power handling capability of electronic capacitors must be considered. It frequently comes up in filter design where the expression "AC ripple" is used. DC power supplies, as we saw earlier, attempt to make pure direct current by filtering out fluctuations. These fluctuations are like ripples on the surface of a pond and represent AC passing through the capacitor. All would be well except that industry cannot build capacitors which have zero resistance.

To calculate power, the expression requires that the AC voltage across the capacitor and the AC current flowing through the capacitor be known. (There might be a DC voltage at the same time, but remember that there cannot be a steady DC current through a capacitor. If there is a pulsating DC, it must be treated like AC.) There is nothing wrong with this expression except its inconvenience. If a capacitor is working in a circuit, it would be relatively easy to measure the voltage across it, and not too difficult to measure the current through it, but designers would like to know ahead of time what is going to happen based upon ratings in catalogs.

The power equation is therefore expressed as:

$$P = E\,I$$

Where: P = Power in Watts
        E = Potential in Volts
        I = Current in Amperes

The first step towards understanding of what is going to happen is Ohm's Law. Ohm's Law is expressed in the following equation:

$$E = I\,R$$

Where: E = Volts
        I = Amperes
        R = Ohms

Take Ohm's Law and substitute IR for E in the power equation, the result allows calculation of power if the current and the resistance is known. The resistance which dissipates heat in capacitors is the ESR for which typical values exist. The only variable to be found is the AC current, I.

$$P = E\,I$$
$$P = (I\,R)\,I$$
$$P = I^2\,R$$

Ohm's Law, which was developed for DC circuits originally, can be plagiarized slightly by substituting impedance Z for R. Alternating current is impeded in three ways, all of which are combined in the expression Z. Ohm's Law for AC circuits may then be expressed as:

$$E = IZ$$

$$I = \frac{E}{Z}$$

Take the new expression for I and return to the power equation and substitute for I.

$$P = I^2 R$$

$$P = \left(\frac{E}{Z}\right)^2 R$$

$$P = \frac{E^2 R}{Z^2}$$

Manufacturers have manipulated themselves into a state where they can write specifications most useful for designers. Manufacturers establish experimentally how much power each physical size of capacitor can handle without getting too hot. (If it gets too hot, its failure rate goes up, and that is bad.) They then consult the catalog impedance curves to get the value of Z, the ESR curves to get the value of R, and calculate the ripple voltage which would be allowed. The expression for ripple voltage is as follows:

$$P = \frac{E^2 R}{Z^2}$$

$$E^2 = \frac{P Z^2}{R}$$

$$E = Z \sqrt{\frac{P}{R}}$$

This AC voltage is known as the "rms" voltage. Fig. 1.26 illustrates the relationship between the peak and rms voltage.

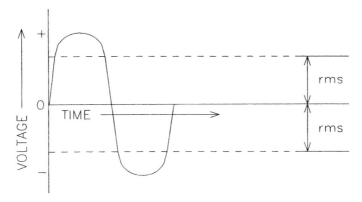

Fig. 1.26 Root Mean Square Voltage

With sine wave AC voltage, the voltage will have some peak voltage but the average voltage will be zero. People can get killed with averages, though, so someone figured out the AC voltage which produces the same heating effect as direct current. This is called the "root mean square" or *rms* voltage. It is equal to the square root of two divided by two ($\sqrt{2}/2$), or if we round off, 0.707 times the peak AC voltage.

The peak voltage is important for another reason. All capacitors have a rated voltage which should not be exceeded by anything—not by DC nor by the peak AC—so the peak value must be calculated as a second restriction in AC applications. Film capacitors and ceramic capacitors are not polar devices. That is, they will work equally well with either positive or negative polarity applied. Electrolytic capacitors, however, are not so flexible, and cannot allow much reverse voltage. If pure AC were applied, of course, the voltage would be in reverse half the time. The answer to this dilemma is called bias voltage.

In Fig. 1.27, both AC and DC voltage are applied to the capacitor, the value of DC being chosen to raise the AC sufficiently above zero to prevent reversal. At the same time, do not raise it too high and exceed the rated voltage with the peak AC.

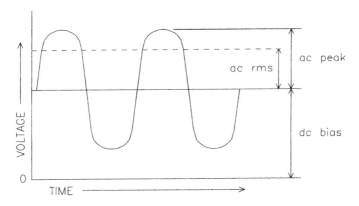

Fig. 1.27   AC Voltage Plus DC Bias Voltage

Application Information

The most important thing a user must decide is which of the numerous types of capacitors will be best for use in the equipment he is designing. Proper selection is the first step in building reliable equipment. To select properly the capacitors to be used, the user must know as much as possible about the types from which he can chose. He should know their advantages and disadvantages; their behavior under various environmental conditions; their construction; and their effect on circuits and the effect of circuits on them; and what makes capacitors fail.

Where more than one type of capacitor may be used in a given application (i.e. molded mica or glass types), consideration should be given to cost and availability (use of strategic materials, multiple sources, etc.).

Capacitors are used as energy storage components to accumulate energy through long periods of time and to discharge the energy over longer or shorter periods. Parallel RC circuits will maintain bias on the grid of a tube for long periods and, as in filter circuits, will smooth out pulsating direct current. Bypass capacitors are used to prevent the flow of direct current without impeding the flow of alternating current and to attenuate low frequency currents while permitting higher frequency currents to pass. In combination with resistors, capacitors are used to reduce radio interference caused by arcing contacts, and to increase the operational life of the contacts.

The temperature at which the dielectric operates is a function of the ambient temperature in which the capacitor is located: the heat which is radiated or conducted to the capacitor; the internal heating of the capacitor due to power losses in the conductors and dielectric; the physical construction and thermal conductivity of the materials inside the capacitors; the transfer of heat internally by conduction and convection to the container; and the heat lost from the container by convection, conduction, and radiation.

The insulation resistance decreases as the temperature increases. The power factor is a complex function of temperature. With polarized dielectrics, temperature-frequency combinations exist where there are large increases in power factor. This may not present any difficulties at low temperatures, since internal heating will raise the dielectric temperature and lower the power factor. An increase in power factor at high temperatures may cause thermal instability and must be considered.

The capacitance of polarized dielectrics is a complex function of the temperature, voltage, and frequency; nonpolarized dielectrics exhibit less change than polarized materials. As the ambient temperature is decreased, many dielectrics will exhibit a large decrease in capacitance with a relatively small change in temperature. The increased power factor at this lower temperature may raise the dielectric temperature sufficiently to recover the lost capacitance. Consideration must be given to the fact that when the capacitor is initially energized at low ambient temperatures, the capacitance will be a small percentage of its nominal value. If the internal heating is effective, the thermal time constant on the capacitor must be considered. A change in the distance between the conductors and the effective area of the conductor due to dimensional changes will cause a change in capacitance.

The dielectric strength of the dielectric material decreases as the temperature increases.

The life of a capacitor, in general, decreases with an increase in ambient temperature. Life as a function of operating temperature is a complex function and should be determined from life-test data. In the absence of this data, the familiar 10°C rule for a chemical reaction may be used as a rough approximation. This rule states that the life decreases by a factor of two for each 10°C rise in temperature. This rule, however, should never be used outside of the temperature range specified by the manufacturer, since chemical reactions of an entirely different nature may take place at extreme temperatures. This rule should not be applied to liquid and gaseous dielectrics without further investigation.

The operating temperature and changes in temperature also affect the mechanical structure in which the dielectric is housed. The terminal seals utilizing elastic materials or gaskets may leak due to the set temperature characteristics. The expansion and contraction of materials with different thermal coefficients may cause leaks at joints. Electrolysis effects in glass terminals increase as the temperature increases. The increase in internal pressure of liquids and gases may cause leaks. A decrease in internal pressure due to the lowering of the temperature may cause internal arc-over.

If the capacitor is operated in the vicinity of a component operating at high temperature, the flashpoint of the impregnant should be taken into consideration.

The dielectric strength of gases is a function of pressure, temperature, frequency, and humidity. Hermetically sealed units must have terminals designed to operate satisfactorily at the required pressure.

The heat loss by convection of a capacitor is a function of pressure and must be considered.

Reduced pressure may produce leaks in hermetically sealed units. An increase in pressure on the container of rolled capacitors in rectangular containers may increase the capacitance by decreasing the distance between the conductors.

The capacitors and mounting brackets, when applicable, must be of a design which will withstand the shock and vibration requirements of the particular application.

Moisture in the dielectric will decrease the dielectric strength, life and insulation resistance, and increase the power factor of the capacitor. In general, capacitors which operate in high humidities should be hermetically sealed. The effect of moisture on pressure contacts which are not gas-tight may result in a high resistance or open contact.

A capacitor may fail when subjected to environmental or operational conditions for which the capacitor was not designed or manufactured. The designer must have a clear picture of the safety factors built into the units, of the safety factors he adds of his own accord, and of the numerous effects of circuit and environmental conditions on the circuit parameters. It is not enough to know only the capacitance and the voltage rating. It is important to know to what extent the capacitance varies with environmental conditions. It also must be recognized that the internal resistance of the capacitor varies with temperature, current, voltage, or frequency; the effects of all of these factors on insulation resistance, breakdown voltage, and other basic capacitor characteristics which are not essential to the circuit but which do invariably accompany the necessary capacitance.

The designer, in selecting a capacitor type for a particular function to be performed, must weigh several factors before a final decision is made. Selection normally starts with the most important characteristic for the application, then selecting and compromising other characteristics.

The capacitance specifications that the circuit designer uses in order to design a circuit which will operate satisfactorily for the desired time requires the following acceptable parameters:
- tolerances according to specification
- capacitance-temperature characteristics
- capacitance-voltage characteristics
- retrace characteristics
- capacitance-frequency characteristics
- dielectric absorption
- capacitance as a function of pressure, vibration, and shock
- capacitor aging in the circuit and during storage.

Capacitance that exists between the capacitor terminals and case may be a consideration, as will stray capacitance and leakage currents. The terminal connected to the outside conductor is often identified by the manufacturer so that the circuit can minimize these effects.

The capacitance-temperature characteristic can be compensated for by using more than one type of capacitor (dielectric) to obtain the required capacitance. The characteristics of other circuit components may also be used for compensation.

The peak voltage which is applied to the capacitor should not exceed the rating on the applicable specification. The safety factor between the peak applied voltage, the test voltage, and the breakdown voltage is of a statistical nature. The same peak voltage, in general, may decrease with:
- aging
- an increase in temperature
- an increase of area of dielectric
- higher frequencies of applied voltage
- a decrease in pressure
- the entrance of moisture into the capacitor.

In many applications, it is necessary to derate the capacitor from the specified voltage to provide the desired performance for the required time. It is to be emphasized that short duration transient voltages cannot be neglected in capacitor applications.

The use of the self-healing properties of certain types of capacitors may not be desirable in circuits where intermittent failures and noise would be troublesome. Some types are not self-healing at low voltages.

Operation of capacitors above the corona-starting voltage will reduce the life and will produce noise. Liquid-impregnated dielectrics have a higher corona-starting voltage than dry solid dielectrics.

When a capacitor is operated at high voltages above ground, and when it is insulated from ground with supplementary insulation, one terminal should be connected to the case, since the division of voltage depends on capacitance between capacitor rolls and case and the capacitance between case and chassis.

The peak charge and discharge currents must be considered on the basis of the time constant of the circuit.

To determine the surface temperature rise of a capacitor, multiply the volt-amperes supplied to the unit by the power factor. This gives the watts lost in the capacitor. Dividing the watts lost by the surface area in square inches will give the approximate surface temperature rise.

Internal heating and ambient temperature must be considered.

Environmental conditions such as corrosive atmospheres, humidity, pressure, fungus growth, shock, and vibration must be considered.

The insulation resistance must be considered, especially when used at higher temperatures.

Balancing resistors should be considered for series operation in DC circuits.

The effective inductance of a large capacitor can be reduced by shunting it with a small capacitor.

The inductance of various types of capacitors varies over wide limits.

Since capacitors have inductance, the operation of capacitors in parallel in circuits with fast rise times or transients may result in transient oscillations.

Poor electrical contacts may open at low voltages and be noisy.

The stored energy in capacitors can be dangerous to personnel and equipment and suitable precautions should be taken.

Extended foil paper capacitors are generally considered superior to inserted tab types, having less inductance and series contact resistance. These are important factors in low voltage applications and in low signal-to-noise ratio circuits.

Oil filled or acid filled capacitors should not be subjected to severe mechanical stresses. Leakage of the fluid can destroy the capacitor together with adjacent components.

Liquid filled units should not be used inverted because internal corona may result.

Nonhermetically sealed capacitors may be pervious to moisture by the process of "breathing."

Capacitors for AC and pulse operation require special ratings and tests.

Trimmer capacitors fall into three categories: multi-turn, single-turn, and compression types. Multi-turn capacitors have either glass, quartz, sapphire, plastic, or air dielectrics, while single-turn devices use ceramic, plastic, or air dielectrics. Compression types use a mica dielectric.

For trimmer capacitor applications requiring low loss, high Q, stability, and tuning sensitivity; glass, quartz, or air dielectric should be selected. Glass and quartz devices are used at frequencies up to 300 MHz. Air dielectrics are usable to about 1 GHz. For frequencies of 1 GHz or above, sapphire dielectrics offer the best performance.

Trimmer capacitors with ceramic and plastic dielectrics are inexpensive, with high grade plastic dielectric devices being usable at frequencies up to 2 GHz.

The most important selection factors are noted below with some of the reasons why these factors are important.

TEMPERATURE EFFECTS:
- Capacitance by variations in dielectric constant or by changing conductor area or spacing.
- Leakage current, through change in specific resistance.
- Breakdown voltage at high temperatures and effect of frequency on heating.
- Current rating, when affected by heating.
- Oil, gas, or electrolyte leakage through seals.

HUMIDITY EFFECTS:
- Leakage current
- Breakdown voltage
- Power factor or Q.

BAROMETRIC PRESSURE EFFECTS:
- Breakdown voltage
- Oil, gas, or electrolyte leakage through seals.

APPLIED VOLTAGE EFFECTS:
- Leakage current
- Heating and its accompanying effects.
- Breakdown of dielectric; effect of frequency.
- Corona
- Insulation to case or chassis.

VIBRATION:
- Capacitance change through mechanical vibration.
- Mechanical distortion of elements, terminals, or case.

CURRENT:
- Effect on internal temperature rise and life of capacitor.
- Ability of conductors to carry currents from a thermal viewpoint.

LIFE:
- Affected by all environmental and circuit conditions.

STABILITY:
- Also affected by all environmental and circuit conditions.

RETRACE:
- After a capacitance change.

SIZE
VOLUME
COST
MOUNTING METHOD

A summary of various capacitor technologies' performance under a neutron radiation field is shown below:

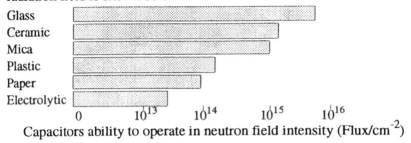

Capacitors ability to operate in neutron field intensity (Flux/cm^{-2})

After a selection of the desired capacitor has been made, reference should be made for a listing of qualified sources.

Chapter 2

Ceramic Capacitors

The value for K comes from the selection of materials and from the geometric arrangement of individual component parts. This chapter covers the dielectric material in ceramic capacitors.

There is one form of ceramic which looks almost exactly like the classical model of a parallel-plate capacitor. A square or circular shaped ceramic dielectric is prepared and coated with conductors on each flat face as shown in Fig. 2.1. If the value of K is known for the dielectric, measure the area of the conductors, the thickness of the dielectric, and directly calculate the capacitance.

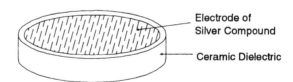

Fig. 2.1 Ceramic Disc Capacitor

In commercial practice, the dielectric is made from finely powered materials, chief of which is barium titanate (K = 1000 to 3000). Disc elements are pressed in dies and then fired at high temperature to produce a very dense structure. Single-plate elements are usually cut from larger sheets of fired ceramic material. Electrodes for both discs and single-plates are formed from a compound containing powered silver, powered glass, and an organic binder. This material is screen printed onto the discs or onto the sheets from which the single plates will be cut. Another firing step removes the binder and melts the glass, binding the silver glass matrix to the ceramic surfaces.

The outer surface is easily solderable, and wires are usually attached in a radial configuration. The hairpin-shaped wires shown in Fig. 2.2 are springy enough to hold the ceramic elements while the assembly is dipped in solder. The lower end of the hairpin is cut off later. This process can be mechanized readily, and dipped discs are among the cheapest capacitors available.

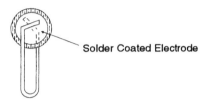

Fig. 2.2   Disc Ceramic with Lead Wires

Unless some special means is taken to remove the electrode compound from the periphery of the single plate element, there is a hazard of conductors bridging across the dielectric to short-circuit the capacitor as seen in Fig. 2.3. Single plates could be printed like discs, but the feeding and locating problems increase cost and reduce accuracy of capacitance achieved.

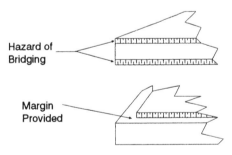

Fig. 2.3   Single Plate Shorting Hazard

A much more sophisticated design is called the "monolithic" ceramic capacitor. It offers much higher capacitance per unit volume. Fig. 2.4 is a cross sectional view and in simplified form. The ceramic material acts both as dielectric and as encapsulant of the basic element. Electrodes are buried within the ceramic and exit only on the ends. The ends are surrounded with a type of powdered silver-glass compound. Fig. 2.4 shows only two electrodes, but 20 or 30 electrodes are very common in commercial practice and 60 or 80 might be used to obtain larger values of capacitance.

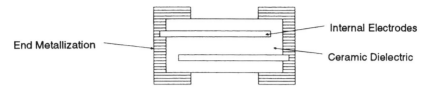

Fig. 2.4 Monolithic Ceramic Element

Fig. 2.5 illustrates the use of three electrodes. The addition of the third electrode has doubled the value of capacitance because of the two layers of the dielectric. The equation for capacitance may be modified by the addition of the term, N, to indicate the number of layers of dielectric in use. The thickness of the layer represents the plate separation, D, in the equation, while the area, A, is the area of dielectric which appears between opposing electrodes. The dimension, L, shown in Fig. 2.5 is representative of this area; the remainder of the electrode length does not face electrodes of opposing polarity, and these portions of the electrodes act only as conductors to the outside world.

$$C = (8.85 \times 10^{-12}) K \frac{A}{D} \times N$$

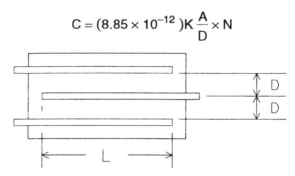

Fig. 2.5 Three Electrode Monolithic Ceramic Element

The manufacturing process for monolithic ceramic capacitors is much more complicated and sophisticated than that needed for discs or single plates. The powered ceramic material is mixed with a binder and cast on moving belts into thin flexible sheets which are wound onto reels and stored. The sheets are then printed with electrode patterns. The "ink" used in this printing is pigmented with finely divided precious metals, usually chosen from among platinum, palladium, and gold. Precious metals are necessary because the electrodes must pass through the firing kiln (above +1000°C) along with the ceramic, and an oxidizing atmosphere must be maintained in the kiln to develop the desired ceramic properties. The use of precious metal electrodes represent a major cost element in making monolithic ceramic capacitors.

After the ink is dried, pieces of the sheet are stacked above one another, each piece representing one dielectric layer. Figure 2.6 shows the electrode patterns are printed so that alternating electrodes exit from opposite ends. Finally, cover layers which do not bear electrodes are placed on top and bottom. The whole assembly is compressed and then fired. During firing, the ceramic sinters together into one homogeneous structure from which we get the name "monolithic."

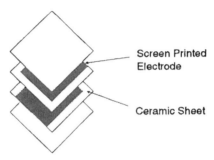

Fig. 2.6 Expanded Monolithic Ceramic Capacitor

The volumetric efficiency of ceramics comes from the high values of K which are possible. This result is in contrast with tantalum dielectrics and other electrolytics which gain efficiency primarily from very close spacing of electrodes. A 50-volt ceramic dielectric, for example, would be about 60 times as thick as a 50-volt tantalum oxide dielectric.

Basic to the ceramic capacitor are the properties of the dielectric materials. There are many dielectric formulations in use to obtain special characteristics of the finished capacitors. In general, stability of capacitance with respect to temperature and voltage are sacrificed when large values of K are sought. While many special formulations are sold, the industry is concentrating on three temperature compensating areas: stable (NP0 or C0G), semistable (X7R), and general purpose (Z5U). The C0G or NP0 is highly stable with respect to temperature and also with respect to voltage and frequency. The others begin to develop erratic deviations in capacitance versus temperature as the value of K goes up. Nevertheless, they are very useful in applications where the temperature changes little.

Multilayer ceramic capacitors are available with a wide range of operating characteristics. The Electronic Industries Association (EIA) and the military have established categories to help divide the basic characteristics into easily specified classes. The basic industry specification for ceramic capacitors is EIA specification RS-198 and as noted in the general section it specifies temperature compensating capacitors as Class 1 capacitors. These are specified by the military under specification MIL-C-20. General purpose capacitors with nonlinear temperature coefficients are called Class 2 capacitors by EIA and are specified by the military under MIL-C-11015 and MIL-C-39014. The high reliability military specification, MIL-C-123 covers both Class 1 and Class 2 dielectrics.

Class 1 - Temperature compensating capacitors are called Class 1 capacitors. They are usually made from mixtures of titanates where barium titanate is normally not a major part of the mix. They have a predictable temperature coefficient (TC) and in general, do not have an aging characteristic. Thus they are the most temperature stable capacitor available. The TC's of Class 1 temperature compensating capacitors are usually NP0 (negative-positive 0 ppm/°C). Other Class 1 extended temperature compensating capacitors are also manufactured in TC's from P100 through N2200 as illustrated on the following page in Fig 2.7.

	TC TOLERANCES[1]			
	Capacitance in pf		Closest MIL-C-20D Equivalent	EIA Designator
	10pf and Over			
	−55°C to +25°C in PPM/°C	+25°C to +85°C in PPM/°C		
NP0	+30, −75	±30	CG	C0G
N030	+30, −80	±30	HG	S1G
N080	+30, −90	±30	LG	U1G
N150	+30, −105	±30	PG	P2G
N220	+30, −120	±30	RG	R2G
N330	+60, −180	±60	SH	S2H
N470	+60, −210	±60	TH	T2H
N750	+120, −340	±120	UJ	U2J
N1500	+250, −670	±250	NONE	P3K
N2200	+500, −1100	±500	NONE	R3L

(1) Indicates the tolerances available on specific temperature characteristics. It may be noted that limits are established on the basis of measurements at +25° C and +85° C and that TC becomes more negative at low temperature. Wider tolerances are required on low capacitance values because of the effects of stray capacitance.

Fig. 2.7 EIA Temperature Compensating Capacitor Codes

Class 2 - General purpose ceramic capacitors are known as Class 2 capacitors and have become extremely popular because of the high capacitance values available in very small size. Class 2 capacitors are "ferro-electric" and vary in capacitance value under the influence of the environmental and electrical operating conditions. Class 2 capacitors are affected by temperature, voltage (both AC and DC), frequency and time. Temperature effects for Class 2 ceramic capacitors are exhibited as nonlinear capacitance changes with temperature.

In specifying capacitance change with temperature for Class 2 materials, EIA expresses the capacitance change over an operating temperature range by a three-symbol code. The first symbol represents the cold end of the temperature range, the second represents the upper limit of the operating temperature range and the third symbol represents the capacitance change allowed over the operating temperature range. A detailed explanation of the EIA system is provided in Fig 2.8.

| EIA CODE ||
| Percent Capacity Change Over Temperature Range ||
RS198	Temperature Range
X7	$-55°C$ to $+125°C$
X5	$-55°C$ to $+85°C$
Y5	$-30°C$ to $+85°C$
Z5	$+10°C$ to $+85°C$
Code	Percent Capacity Change
D	±3.3%
E	±4.7%
F	±7.5%
P	±10%
R	±15%
S	±20%
T	+22%, −33%
U	+22%, −56%
V	+22%, −82%

Fig. 2.8 EIA Temp. Stable and General Codes

Effects of Voltage - Variations in voltage affects only the capacitance and dissipation factor. The application of DC voltage reduces both the capacitance and dissipation factor while the application of an AC voltage within a reasonable range tends to increase both capacitance and dissipation factor readings. If a high enough AC voltage is applied, eventually it will reduce capacitance just as a DC voltage will.

Capacitor specifications specify the AC voltage at which to measure (normally 0.5 or 1V AC) and application of the wrong voltage can cause spurious readings. Applications of different frequencies will affect the percentage changes versus voltages.

Effects of Frequency - Frequency affects capacitances and dissipation factor. Variation of impedance with frequency is a very important consideration for decoupling capacitor applications. Lead length, lead configuration and body size all affect the impedance level over more than ceramic formulation variations.

Effects of Time - Class 2 ceramic capacitors change capacitance and dissipation factor with time as well as temperature, frequency, and voltage. This change with time is known as aging. Aging is caused by a gradual realignment of the crystalline structure of the ceramic and produces an exponential loss in capacitance and decrease in dissipation factor versus time.

If a ceramic capacitor that has been sitting on the shelf for a period of time is heated above its curie point, (+125°C for 4 hours or +150°C for 1/2 hour will suffice) the part will de-age and return to its initial capacitance and dissipation factor readings. Immediately after de-aging, the capacitance changes rapidly. The basic capacitance measurements are normally referred to a time period sometime after the de-aging process. Various manufacturers use different time bases but the most popular one is one day or twenty-four hours after "last heat." Changes in the aging curve can be caused by the application of voltage and other stresses. Heating may cause changes in capacitance due to de-aging. This is why MIL specifications allow capacitance changes after testing, such as temperature cycling, moisture resistance, etc. The application of high voltages such as dielectric withstanding voltages also tends to de-age capacitors and is the reason why rereading of capacitance after 12 or 24 hours is allowed in MIL specifications after dielectric strength tests have been performed.

Effects of Mechanical Stress - High K dielectric ceramic capacitors exhibit some low level piezoelectric reactions under mechanical stress. As a general statement, when the piezoelectric output is high, the higher the dielectric constant of the ceramic. It is desirable to investigate this effect before using high K dielectrics as coupling capacitors in extremely low level applications.

Reliability - Historically, ceramic capacitors have been one of the most reliable types of capacitors in use today. The approximate formula for the reliability of a ceramic capacitor is:

$$\frac{L_o}{L_t} = \left(\frac{V_t}{V_o}\right)^x \left(\frac{T_t}{T_o}\right)^y$$

where:
L_o = operating life
L_t = test life
V_t = test voltage
V_o = operating voltage
T_t = test temperature in °C
T_o = operating temperature in °C
X, Y = Historically for ceramic capacitors:
   Exponent X has been considered as 3.
   Exponent Y for temperature effects typically tends to run about 8.

Application Information

Ceramic capacitors are primarily designed for use where a small physical size with comparatively large electrical capacitance and high insulation resistance is required. Ceramic capacitors are substantially smaller than paper or mica units of the same capacitance and voltage rating. Ceramics can be used where mica or paper capacitors have too wide a capacitance tolerance. The lead placement makes ceramic capacitors suitable for printed circuit use.

General purpose ceramic capacitors are not intended for precision applications but are suitable for use as bypass, filter, and noncritical coupling elements in high frequency circuits where appreciable changes in capacitance, caused by temperature variations, can be tolerated. These capacitors are not recommended for use directly in frequency determining circuits.

Typical applications for general purpose ceramic capacitors include resistive-capacitive coupling for audio and radio frequency, RF and intermediate frequency cathode bypass, tone compensation, automatic volume control filtering, volume control RF bypass, antenna coupling, and audio-plate RF bypass. All of these applications are of the type where dissipation factor is not a critical factor, and moderate changes due to temperature, voltage, and frequency variations do not affect the proper functions of the circuit.

Because the ceramic dielectric capacitor materials have molecular polar moments, the dielectric constants of some mixes reach hundreds (even thousands) of times the value of paper, mica, and plastic films. This results in ceramic dielectric capacitors having the largest capacitance-to-size ratios of all high-resistance dielectrics.

Temperature compensating capacitors are designed for use primarily where compensation is needed to counteract reactive changes, caused by temperature variations, in other circuit components. However, they can be used in any precision-type circuit where their characteristics are suitable.

The temperature compensating capacitors are recommended for use in frequency determining circuits. Typical applications include oscillator, radio frequency (RF), and intermediate frequency (IF) circuits. Frequency drift due to temperature effects can be compensated individually in each circuit.

In IF stages where the frequency variation is uniform, satisfactory operation can be obtained by designing the temperature-compensating capacitor into the oscillator circuit. RF circuit reactive changes caused by temperature variations cannot be compensated for in the oscillator circuit. Where more critical tuning accuracy is required, it is necessary that compensating capacitors be inserted directly into each circuit.

In RF circuits tuned by a variable capacitor, a shunt compensating capacitor of low value and high compensating characteristics may be used.

In slug-tuned circuits, the total capacitance required can be provided by using a compensating capacitor having the desired temperature coefficient.

In oscillator circuits, more linear tuning can be obtained by using proper temperature coefficients in both the series and the shunt capacitances of the tank circuit.

If possible, the temperature/time curve of the selected capacitor should be the exact opposite of the temperature/time curve of the coil (or other component) being stabilized. Combinations of different capacitance values and temperature coefficients provide the designer more precise compensation than can be obtained from a single capacitor.

Disk and thin-plated subminiature types are extremely compact and have an inherent low-series inductance due to their construction. The placement of the leads facilitates making close-coupled low inductance connections and are suitable for printed circuit applications.

High insulation resistance allows usage in electron vacuum-tube plate and grid circuits. Their extremely low leakage and small physical size make them suitable for transistor circuit design. They are also useful in filter and bypass circuits.

During circuit design, consideration should be given to the changes in dielectric constant caused by temperature, applied frequency, electric field intensity, and shelf aging. Another consideration should be given to the physical placement of compensating (and compensated for) components. Locations near hot transistors will cause much greater reactive variations than spots adjacent to a cool, external chassis.

The dielectric constant exhibits a considerable dependence on field strength. Large variations in capacitance may be experienced with changes in AC or DC voltages. The dielectric constant may decrease with time and may be as low as 75 percent of the original value. The dielectric constant is dependent on frequency and decreased as the frequency is increased; it also decreases with temperature.

Ceramic dielectrics are frequency sensitive; both the capacitance and the capacitance change with temperature will be different at different measuring frequencies. For extremely accurate compensation, the capacitors should be measured at the proposed operating frequency.

Ceramic dielectric materials are nonhygroscopic, effectively impermeable, and have practically no moisture absorption even after considerable exposure to humid conditions. Thus, these capacitors are intended to operate, through their full temperature range, at relative humidities up to 95 percent. Nevertheless, the termination materials under moisture conditions are subject to ionic migration which can cause capacitor failure.

When the silver electrodes in the ceramic capacitor are exposed to high humidities and high DC potentials, silver ion migration may take place and short-circuit capacitors after relatively short periods of time. During periods of storage, excessive moisture should be avoided since the encapsulation material may absorb moisture and silver ion migration may occur when the capacitors are later put into service.

Care should be used in soldering the leads. Excessive heat may damage the encapsulation and weaken the electrode to terminal contact. Sudden changes in temperature, such as those experienced in soldering, can crack the encapsulation or the ceramic dielectric. Leads should not be bent close to the case nor should any strain be imposed on the capacitor body to avoid fracturing the encapsulation of the ceramic dielectric.

For the recommended applications, the dissipation factor is negligibly low. The power factor decreases as temperature is increased; this provides an advantage where operation above room temperature is required.

Capacitor aging is a term used to describe the negative, logarithmic capacitance change that takes place in ceramic capacitors with time. The more stable dielectrics have the lowest aging rates, however, all ceramic capacitors with high dielectric constants display an aging characteristic. General purpose dielectrics comprise this high dielectric constant family.

High K ceramic dielectrics with a barium titanate formula exhibit the phenomenon known as Curie Point crystal-phase transformation. Simply stated, most of the tiny crystals that make up the ceramic microstructure are of cubic symmetry at a temperature of +120°C and above. Below +120°C, these same crystals take on a tetragonal shape. The specific relationship between this crystal-phase transformation and aging is not clearly understood, but it is known that they are directly related. As the crystals change from cubic to tetragonal shape, stresses are set up in the dielectric and are subsequently relieved gradually. This electrical "aging" phenomenon seems to follow the same logarithmic patterns observed in mechanical models of stress relief. Each time the capacitor is heated to approximately +120°C (Curie Point), all of the negative capacitance change that may have taken place is recovered. Upon cooling, the aging cycle begins again. This recovery process is commonly referred to as "de-aging." The entire process of aging and de-aging is predictable and can be repeated infinitely.

Another important factor that affects capacitor aging is the application of a polarizing voltage. The application of a DC voltage approximately equal to the capacitor's rating will cause an abrupt negative capacitance change; however, when the voltage is removed, the capacitor does not return to its original polarized value. If this exercise were performed on a capacitor with a known aging characteristic and the results were plotted, the resultant curves would show that capacitance change is negative and logarithmic with respect to time. The application of DC bias voltages and subsequent dielectric polarization of the capacitor micro-structure serve to relieve some of the stresses in the dielectric. This will move a point on the aging curve forward in time.

Most general purpose state-of-the-art dielectrics found in industry have aging rates varying from 1.5 percent to 4 percent.

The following points should be kept in mind when you are dealing with the phenomenon of ceramic capacitor aging:
- The process is completely repeatable and predictable.
- Capacitance change is negative and logarithmic with respect to time.
- Application of DC bias can move a point on the curve forward in time.

This wide capacitance change, as a result of "shelf" aging and temperature cycling, is why tight-tolerance, high K ceramics are not common in the electronics industry.

The mixes of different temperature coefficients are made by varying the percentages of high K dielectrics (such as titanium dioxide) in the low-loss ceramic. The temperature becomes increasingly more negative with the increase in dielectric constant. For example:

Material	K	Temperature Coefficient
Titanium dioxide	85	-750
Low-loss ceramic	6	+100

As a result, for any given size of capacitor, the relative capacitance will be high with a high negative temperature coefficient, and vice versa. With present day manufacturing methods, a high degree of consistency and reproducibility is obtained for the different coefficients.

Temperature coefficients are not linear with respect to temperature. Measurements taken at +25°C and +85°C will show a change of value which, when divided by 60 (the temperature differential), does not represent the change in capacitance to be expected for each degree change in temperature. The coefficient is therefore not expressible by a single number.

It is recommended that supplementary insulation be used where the capacitor body will normally contact parts with a potential difference of more than 750 volts.

Ceramic chip capacitors are intended to be used in surface mount, and in thin or thick film hybrid circuits.

Designers are cautioned to give consideration to the change in dielectric constant with temperature, shelf aging, and electric-field intensity. Also, they should recognize that the insulation resistance may vary with humidity and organic contamination of the ceramic chip surfaces.

It should be noted that when pure silver is used for the terminations, silver migration between the terminations may occur under conditions of simultaneous application of high humidity and DC voltage. This produces a troublesome electrical leakage path across the capacitor chip. Addition of about 20 percent of palladium to the silver to form an alloy will retard the tendency toward silver migration. Complete overcoating of the silver termination by the lead-tin bonding solder also will retard the tendency toward silver migration. Addition of about 3 percent of silver to the lead-tin bonding solder will tend to reduce the leaching of the silver from a silver termination during the solder bonding operation.

Care should be taken in the selection of the capacitor chip mounting substrate material. Voltage temperature limits, resistance to thermal shock, and reliability may be affected as a result of mounting on substrates with dissimilar coefficients of expansion from the capacitor material.

Variable ceramic capacitors are small sized trimmer capacitors designed for use where fine tuning adjustments are periodically required during the life of the equipment. Normally they are used for trimming and coupling in such circuits as intermediate frequency, radio frequency, oscillator, phase shifter, and discriminator stages. Because of their low mass, these capacitors are relatively stable against shock and vibration which tend to cause changes in capacitance. Where a higher order of stability is required, air trimmers should be used. The minimum rated capacitance of the variable capacitor should not be greater than the minimum design value specified; however, the minimum capacitance value may be less than the minimum value specified. The maximum capacitance should not be less than that specified and not greater than 50 percent more than the maximum value specified. Capacitance and adjustment are relatively linear.

Variable ceramic capacitor's principle of operation is similar to that of an air-dielectric tuning capacitor where the overlap of the stator and rotor determines the capacitance; however, the ceramic dielectric replaces the air dielectric. Rotors may be rotated continuously; full capacitance change occurs during each rotation.

Ceramic transmitting capacitors are designed primarily for use in applications where very high RF currents, high KVA ratings and high working voltages exist. Special capacitor design reduces terminal self-inductance to permit higher frequency operation. A humidity resistant silicone coating and long flashover path assures safe, high voltage operation in free air. Stable temperature-capacitance characteristics and rugged mechanical construction is required for maximum reliability.

Ceramic transmitting capacitors are extensively used in transmitters, antennas, induction heaters, X-ray, diathermy and electronic welding equipment.

Ceramic spark-gap capacitors provide a safe, reliable discharge path for stray transient over-voltages and static voltage build-up for many industrial and commercial equipment applications. A typical application in color TV utilizes a minimum capacitance capacitor which is inserted between the grid lead and chassis ground. This protects the control components by providing a low impedance path to ground for transient voltages of 1500 volts and above.

Across-the-line ceramic disc capacitors should be safety agency approved (UL, CSA, etc.). They must comply with the appropriate standard for the following applications:
- Across-the-Line
- Antenna Isolation
- Line-Bypass.

CERAMIC CAPACITOR COMPARISION CHART

EIA Class and Ceramic Materials	Application	Features	Benefits
Class I Temperature compensating & temp. stable (disc and monolithic) NPO, N080, N150, N330, N470, N750, N1500, N3300.	• Impedance matching in frequency coupling. • Circuit drift stabilization in timing circuits. • Communictions tuners & active filters. • LC and RC tuned circuits and filters.	• NPO has no capacitance drift with wide temp. changes. • Tight tolerance available. • Low dissipation factor. • Temperature compensation with linear negative T.C.	• Stable tuned circuits. • Extremely low frequency drift. • Hi-Q for "notch filters." • Low capacitor change with applied voltage stabilizes frequency drift.
Class II General Purpose (disc and monolithic) X5P, Y5E, X5F, Z5P, Z5U, X5U, Z5V, X7R.	• Generally applicable for all bypass, coupling, blocking, and filtering requirements.	• Nonpolar construction. • Good high frequency performance. • Selection of temperature characteristics. • High insulation resistance at all specified temps.	• Simple circuitry. • High volumetric efficiency. • Temp. stability matched to applications. • Highly reliable in high temp. applications.
Class III Semiconductor type Y5P and other values.	• Applicable for high density mounting on circuit boards where bulk capacitance is required.	• Nonpolar construction. • High bulk capacitance. • Stable temp. characteristics.	• Simple circuitry. • Highest volumetric efficiency available in disc types.

Chapter 3

Plastic Film Capacitors

The original film capacitors did not use plastic film at all, but paper. The pores in the paper, and also various chemical and physical contaminants, all mitigated against very compact or reliable capacitors. The next step was to impregnate the paper with some dielectric fluid such as an oil. This really produced a mixed dielectric with the characteristics, such as the value of K, the breakdown voltage, and temperature stability, being comprised of contributions from both paper and the impregnant. Such capacitors still have advantages and are used in some electric as opposed to electronic applications. As the various synthetic plastic materials— particularly the thermoplastic materials—were developed, their superiority over paper for most applications became apparent. Usually, only one type of plastic film is used in any given capacitor, although mixtures of two different plastics, or plastic and paper, or plastic and impregnated paper, are all possibilities.

It is not difficult to move from the parallel-plate model to the design of what is called the "film-foil" capacitor. The dielectric material is sandwiched between two pieces of metal foil which become the electrodes. The thickness of the plastic film determines the separation between electrodes, and the operating area of the capacitor is the area of the electrodes opposing each other. To contain this structure in a practical space, the sandwich is wound into a jelly roll and then tightly anchored. Frequently, more than one piece of plastic film will be used to comprise the dielectric because there are always chances of pinholes in the plastic. The odds against having two pinholes line up opposite one another are very small.

Once the jelly roll is wound up, the next step is connecting wires to the two electrodes. One method is to wind the electrodes exactly opposite each other with a margin of dielectric for safety in the edges. At one or more points in the jelly roll, a metal tab would be inserted and bonded by pressure as shown in Fig. 3.1.

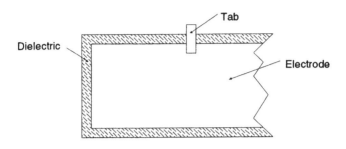

Fig. 3.1   Film-Foil With Tab

All the current flowing through the capacitor must be collected at the one point of the electrode which is touching the tab. The tab would be located about halfway down the length of the electrode to minimize the difference in conducting path over the entire electrode length. Both resistance and inductance in this design will be higher than in the "extended foil" design of Fig. 3.2.

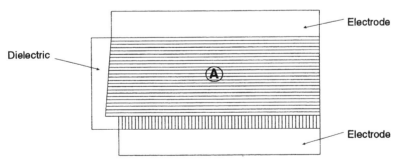

Fig. 3.2   Overlap In Extended Foil Design

The foils overlap the dielectric on opposite edges. This technique reduces the effective area, which is limited to that denoted "A" in Fig. 3.2. Only the area of an electrode which is opposed by the other electrode is effective. After the jelly roll is wound up, it is possible to gain connection the the entire length of each electrode. This connection is usually made to the electrode edges by a special solder or sometimes by sprayed molten metal. The electrodes are usually made of aluminum foil because of its good conductivity and relatively low cost.

The present drive towards miniaturization, closer electrical tolerances, and higher operating temperatures is being met by the use of thin plastic film dielectrics in the construction of capacitors. The greatest advantage of plastic film dielectrics over natural dielectrics (such as paper and mica) is that the plastic film is a synthetic that can be made to meet specific requirements, such as thickness of dielectric and high heat resistance. Many plastic film capacitors are not impregnated but are wound and encased "dry." Plastic dielectric capacitors have insulation resistance values far in excess of those for paper capacitors and since they are nonabsorbent, their moisture characteristics are superior to those of mica.

There are several types of plastic films available for use as a capacitor dielectric. They may be used individually or in a combination with other films in order to obtain the compromised advantages of the specific electrical characteristics of each individual film. The more common films include polyethylene terephthalate and polycarbonate. When properly applied, plastic dielectric films lead to the solution of many special capacitor problems.

Capacitors using polyethylene terephthalate as the dielectric are perhaps the most common of the plastic film types on the market today. Some manufacturers use only one sheet of plastic film for those with low voltage ratings, whereas, at least two sheets of paper are used in conventional paper types. The principal advantage of polyethylene terephthalate dielectric capacitors is the higher order of insulation resistance values available over the dielectric's temperature range. Polyethylene terephthalate dielectric capacitors have an insulation resistance that is normally about 100,000 MΩ/μF at room temperature and about 25,000 MΩ/μF at +85°C. These insulation resistance values decrease considerably when polyethylene terephthalate dielectric capacitors are impregnated. However, a higher volt per mil rating is made possible by impregnation and the possibility of corona and catastrophic failures due to pinholes in the dielectric are minimized.

Another major class of plastic film capacitors is known as "metallized film," which is shown in Fig. 3.3. This design is much newer and is growing fast because it offers much higher volumetric efficiency. Instead of using free-standing metal foil, the electrodes are vacuum-deposited on the dielectric film, and their thickness might be 1/100 of that of foil electrodes. The thinner electrodes save space, and the resistance of the metallized electrode is correspondingly higher.

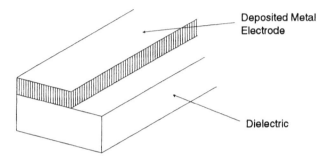

Fig. 3.3 Metallized Film

Fig. 3.4 shows how resistance and inductance are minimized by using an analog to the extended foil design.

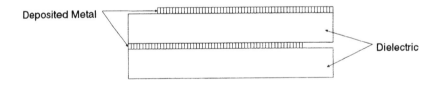

Fig. 3.4 Metallized Areas Extending to Opposite Edges

A margin of un-metallized dielectric remains when the electrode is deposited. Each piece of film is rolled together with a corresponding piece having its margin on the opposite side. Connection to the edges of the metallization by soldering is impractical, and only the sprayed-metal technique is used.

The construction of metallized plastic capacitors shown in Fig. 3.4 differs from conventional plastic capacitors. Instead of having separate layers of metal foil (capacitor plates) and plastic dielectric, the metal comprising the capacitor plates is imposed directly on one side of the plastic dielectric by means of a metallizing process. This technique results in an overall size reduction for metallized plastic capacitors when compared to conventional plastic-foil capacitor types of equal rating. This space saving is the outstanding feature of the metallized plastic capacitor.

The Fig. 3.5 illustrates another advantage resulting from the metallizing technique. The metallized capacitors have a self-healing characteristic called "clearing." The metallic film imposed on the plastic is very thin and if a breakdown by either a hole or contaminant occurs, a tiny area of the thin film surrounding the breakdown point burns away. This leaves the capacitor operable, but with a slightly reduced capacitance. In the conventional plastic-foil type (where the foil is thicker), sustained conduction can occur on a breakdown causing a large area of the plastic surrounding the breakdown to be carbonized resulting in a permanent short-circuit.

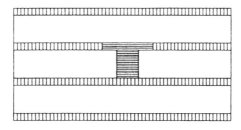

Fig. 3.5 Cleared Fault

The breakdown of the metallized plastic capacitor can be either of two types; (1) a complete breakdown lasting for only a moment (momentary breakdown) or (2) a sharp reduction in insulation resistance lasting for an extended period of time, but eventually returning to normal (period of low insulation). The general characteristics of the metallized plastic type, aside from the breakdowns, are similar to the conventional plastic type except for a significantly lower insulation resistance, approximately in the order of 10 to 1. The self-healing characteristic has the good features of increasing yield if all the faults can be removed during the manufacturer's in-plant burn-in. It may seem to be much better if the fault did not cause trouble until the capacitor were in use. There would only be a momentary flow of current and we would have a nearly whole capacitor again. In analog circuits, this probably is true most of the time. In digital circuits, however, it is very possible that the flow of current during the clearing action would cause a spurious signal and upset the logic of the circuit. Manufacturers have spent a considerable amount of time in determining rates of clearings as they affect reliability.

Advances in technology have now allowed depositing plastic onto metal instead of depositing metal onto plastic. With this process, both sides of the foil are coated resulting in two layers of dielectric to guard against pinholes in the formed jelly roll. Also, pinholes are extremely rare because of the uncanny knack of the dielectric to find surfaces upon which to grow. The dielectric can be deposited in very thin layers resulting in very thin capacitors.

The present drive towards miniaturization, closer electrical tolerances, and higher operating temperatures is being met by the use of thin plastic film dielectrics in the construction of capacitors. The greatest advantage of plastic film dielectrics over natural dielectrics (such as paper and mica) is that the plastic film is a synthetic that can be made to meet specific requirements (such as thickness of dielectric and high heat resistance).

Many plastic film capacitors are not impregnated but are wound and encased "dry." Plastic dielectric capacitors have insulation resistance values far in excess of those for paper capacitors and since they are nonabsorbent, their moisture characteristics are superior to those of mica.

There are several types of plastic films available for use as a capacitor dielectric. They may be used individually or in a combination with other films and with paper in order to obtain the compromised advantages of the specific electrical characteristics of each individual film. The more common films include polyethylene terephthalate and polycarbonate. When properly applied, plastic dielectric films lead to the solution of many special capacitor problems.

Application Information

Plastic Film Capacitors

Plastic film capacitors are designed for use in circuit applications that require high insulation resistance, low dielectric absorption, low loss factor over the wide temperature ranges, and where the AC component of the impressed voltage is small with respect to the DC voltage rating.

NOTE: These capacitors may be used where an AC component is present provided that:
- The sum of the DC voltage and the peak AC voltage does not exceed the DC voltage rating.
- The peak AC voltage does not exceed 20 percent of the DC voltage rating at 60 HZ; 15 percent at 120 HZ; or 1 percent at 10,000 HZ.

Where heavy transient or pulse currents are encountered, allowance must therefore be made in the selection of a capacitor.

All metal parts, fittings, conductors, and attachments which operate at higher potential than other adjacent parts of the capacitor housing should be carefully finished in order to insure that sharp corners and edges are removed to minimize the possibility of corona discharge. Parts from which the removal of sharp corners and edges would be impractical, such as conductors, should be spaced in such a manner as to prevent harmful corona discharges.

Film capacitors are suited for filters, multivibrator timing capacitors, A/D converters, integrators, and other applications where capacitance stability, which is obtained by locking the capacitor body in place with a solid impregnant, is essential.

Film capacitors have many outstanding electrical characteristics and excellent volumetric efficiencies, up to 20 times greater than mica, glass and porcelain.

Polyethylene terephthalate (polyester) capacitors are intended for high-temperature applications similar to those served by hermetically sealed paper capacitors, but where high insulation resistance at the upper temperature limits is required.

Paper and polyethylene terephthalate capacitors are intended for use in applications where high insulation resistance is necessary.

Polycarbonate capacitors are intended for applications where minimum capacitance changes with temperature are required; these capacitors are especially suitable for use in tuned and precision timing circuits.

Polystyrene film and foil capacitors are for use in LC filters, particularly in telephone equipment where high quality requirements are imposed on precision, stability, humidity, dissipation factor, and reliability.

Polypropylene film and foil capacitors are for use in tuned circuits, filter networks, and timing circuits where precision, low losses, and reliability are of prime importance.

Polypropylene film and foil capacitors consist of a series constructed low inductance wound cell of polypropylene film, aluminum foil and double metallized polyester film. These capacitors are for applications where high currents and steep pulses occur. They are commonly used for deflection circuits in television receivers, to operate at line frequency with high peak currents.

Plastic film trimmer capacitors permits the calibration and adjustment of capacitance. Typical dielectrics are polypropylene, polyethylene, or polycarbonate. For high temperature and precision applications, a PTFE Teflon® dielectric material is used.

Metallized Film Capacitors

Metallized polycarbonate film capacitors are designed for circuits that require stable performance characteristics over a wide temperature range, small size, and operational reliability.

Metallized polycarbonate film capacitors are primarily intended for use in circuit applications which require nonpolar behavior, high insulation resistance, low capacitance change with temperature, low capacitance drift over the temperature range, and low dielectric absorption.

Metallized film capacitors are rated for continuous AC operation, in addition to continuous DC operation under conditions where the AC component of voltage is small with respect to the DC voltage. These capacitors can exhibit periods of low insulation resistance and should only be used in circuits that can tolerate occasional momentary breakdowns. They should not be used in high impedance low voltage applications.

Another advantage resulting from the metallizing technique is that the capacitors are self-healing.

Metallized polycarbonate film capacitors are primarily intended for use in power supply filter circuits, bypass applications, SCR commutating, timing, and 400 Hz AC power circuits.

Metallized polyester (polyethylene terephthalate) film capacitors are used for general purpose and industrial use in electronic equipment requiring coupling and decoupling applications.

Metallized polyester and paper capacitors are used for interference suppression in small household appliances (i.e., coffee grinders or mixers,) audio and TV circuits, and in general industrial applications (i.e., test and measuring equipment.)

Dry metallized polypropylene capacitors are for AC applications which require operation in symmetrical sinewave circuits at 60 Hz or lower.

FILM CAPACITORS APPLICATIONS GUIDE

	General Purpose	Tight Tolerance ±1%, ±2% Flat T.C.	Pulse High Current/VAC RMS	Suppression Protection Frequency
	(ALL)	(Polyprop./ Polysty)	(Polyprop./ Metallized)	(Polyprop./ Metallized)
Blocking	X	X	X	
Coupling/ Decoupling	X	X	X	X
Bypassing	X	X	X	X
Filtering				X
Energy Storage	F/F		X	X
Timing	X	X	X	
Smoothing	X	X	X	X
Tuning	X	X	X	X
Frequency Discrimination		X		X
Pulse Discharge	F/F		X	X
Arc Suppressing	F/F		X	X
Transient Voltage Suppression	F/F		X	X

F/F = Film/Foil Only

Chapter 4

Aluminum Electrolytic Capacitors

The ability of a capacitor to store electrical energy is a direct function of its mechanical geometry and its chemical composition. The amount of energy that it can store is given by the equation:

$$Q = CV$$

Where: Q = the magnitude of the stored charge in coulombs
C = the capacitance in farads
V = the applied voltage

The capacitance is determined by the equation:

$$C = (K)\frac{A}{D}$$

Where: K = dielectric constant of the material separating plates
A = directly opposing area of the plates
D = distance between plates

With this equation, the units are: capacitance in farads, the area (A) in square meters, and the distance between electrodes (D) in meters. K is simply a ratio and a pure number without dimensions. When units other than farads and meters are used, different constants are used, i.e., microfarads and inches.

Engineers and scientists have been wrestling with these factors for generations, in the perennial effort to pack more and more capacitance into less and less space, in conformance with the unending trend of equipment miniaturization.

Obviously, increasing the area (A) of the capacitor plates will increase the capacitance of the device. This would tend to increase size, but since only the area, not the thickness, of the plates is significant (in most applications) the plates could be made thinner to offset the increase. Ways were developed to produce ever thinner metal foils, and to deposit thin metallic films directly on both sides of a paper or plastic ribbon which then can be rolled up.

Reducing the thickness of the dielectric separator will also increase the capacitance of the device by reducing the distance (D) between the plates; this also reduces the size for a given capacitance, or allows more capacitance to be installed in a given space. Advancements in the production of high quality, homogeneous plastic films of very thin gauges have enabled substantial reduction in capacitor size, combined with worthwhile increase in capacitance per unit volume.

Use of dielectric materials having higher dielectric constants (K) will also increase the capacitance of the device, hopefully with a decrease (or at least no increase) in unit size. The search for better materials will continue.

One of the major breakthroughs in this field occurred in the early 1900's; the development of the electrolytic capacitor, a brilliantly ingenious expedient for obtaining high capacitance in a small space. Essentially, it consisted of an aluminum foil ribbon, on the surface of which a thin film of aluminum oxide has been formed electro-chemically, and a water-based electrolyte fluid which acts as the opposing plate. The oxide-coated foil, a second strip of aluminum foil, and a porous strip of paper interposed between them were rolled up together, and suspended in the liquid electrolyte, which penetrated the porous spacer.

The physical relationship as described in the previous paragraph is diagrammed in Fig. 4.1.

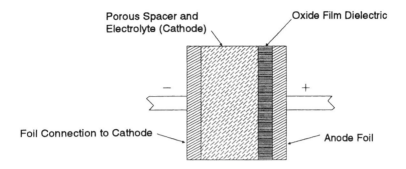

Fig. 4.1   Polarized Electrolytic Capacitor

The oxide-coated foil is the positive plate (anode), the aluminum oxide film is the dielectric, and the fluid electrolyte is the negative plate (cathode). The second strip of aluminum foil serves only as a connection in broad and intimate contact with the negative plate electrolyte, and is usually bonded to the aluminum can that houses the capacitor. The porous strip prevents direct short circuits between the two foil strips.

The oxide dielectric has a thickness on the order of 0.01 micron; thus the distance (D) between the "plates" has been reduced almost to the vanishing point. Furthermore, the dielectric constant (K) of this oxide is approximately 8, compared to 5 for paper, or 3 for polycarbonate film. As a result, the capacitance per cubic inch in an electrolytic capacitor is increased tremendously, compared to conventional capacitor designs, even in the original electrolytic versions. Over the years since their inception, there have been continuous improvements in electrolytic capacitor designs, and advancements in their technology. One of the most significant was that of etching the anode plate. The etching action exposes the grain structure, enormously increasing the area of the surface for a given area of foil. Etching of a given area of aluminum foil results in a many fold increase in the actual surface area facing the electrolyte plate, in the finished capacitor. Two major advancements were the development of non-aqueous and solid electrolytes, and of practical manufacturing techniques for the production of high-purity aluminum foil. Both of these factors will be examined in detail later, in conjunction with capacitor fabrication.

While the energy storage capabilities of the aluminum electrolytic capacitors are impressive, electrolytic construction has certain inherent limitations that affect the use and performance of these capacitors. Safe operating voltages are limited to about 450 Volts. The oxide dielectric has rectifier properties, blocking current flow in one direction but offering low resistance in the opposite direction; it is therefore limited to DC applications, and a voltage reversal of more than a volt or two will cause breakdown of the film and destruction on the capacitor. (Nonpolarized types for AC applications are available. Their construction is essentially the same as shown is Fig. 4.1, but both foils are coated with oxide dielectric, constituting two capacitors connected back to back.) The power factor of electrolytics is considerably higher than those of other capacitor types, and the broad plate area makes for appreciable leakage.

Production Technology

The design and fabrication of an electrolytic capacitor is a complex science. The physical principles involved, and their interlocking—and occasionally mutually exclusive—relationships have been the subject of continuous, heavy research for more than 80 years. Some of the major developments that have evolved are discussed below.

The Anode (Positive Plate)

Capacitance is directly proportional to the surface area of the capacitor plates. This factor has been uniquely exploited in electrolytic design by etching the surface of the anode foil. This is accomplished either chemically, by immersion in an acid bath such as hydrochloric acid, or electrochemically by immersion in a conductive, corrosive bath such as a solution of sodium chloride, and applying an electric current to the foil and solution. In both cases, the etching action exposes the grain structure of the metal, enormously increasing the area of the surface for a given area of foil. The degree of etch is controlled by immersion time in chemical etching, and by the regulation of current flow in electrochemical etching.

The presence of impurities (principally copper, silicon, magnesium, iron, and zinc) in the anode foil can result in early failure of an electrolytic capacitor. Particles of other metals do not form an oxide barrier layer as aluminum does, hence constitute leakage paths. They also form galvanic couples with aluminum and will produce hydrogen gas in the presence of an electrolyte, besides reducing the efficiency of the oxide-layer barrier and causing generation of excessive heat. For these reasons, high-purity aluminum is used for foils, and the electro-chemical etching process is the most suitable for use on this material. Lower purity aluminum is used for the cathode foil.

In the electrochemical etching process, high etching-current density produces a fine etch pattern, with a very high surface gain. However, when this surface is anodized (as discussed later) to voltages above 100 Volts, the thickness of the oxide layer formed will bridge over some of the fine depressions of the etch pattern, reducing the surface gain, and the forming reaction will cause mechanical erosion of the peaks of the etch, further reducing the effective area.

Using the lower etching-current density produces a coarser surface, but the higher resistance to erosion and over-bridging results in a higher final capacity, when anodized at higher voltages. Thus, there is a trade-off relationship between etch coarseness and forming voltage in achieving maximum capacity at a selected working voltage, in a capacitor of given size. It is possible to custom tailor the etch for optimum capacity at a given voltage.

The foil is run as a continuous ribbon through the pre-cleaning bath, then over a roller which supplies the current for etching, then down into the etching tank, then through a series of baths that remove the etch solution, neutralize residual salt, remove loose metal particles, and wash away any remaining materials carried over from the processing tanks. The foil is then dried and immediately rolled, and protected from the atmosphere to prevent formation of "nonbarrier" oxide prior to its entry into the anodizing, or forming, process.

Two types of "anodized" films can be formed on aluminum. In contact with moist air, the surface layer of aluminum forms a porous oxide of regular structure and low resistance, known as nonbarrier oxide. When immersed in certain electrolyte solutions and connected to a DC power source as an anode with the solution as a cathode, the surface layer of aluminum forms an impervious, amorphous film of aluminum oxide having the property of restricting the flow of current in one direction and permitting it to flow in the opposite direction. This barrier oxide layer has a thickness which is a function of the applied voltage—about 14 angstroms (Å) per volt—at room temperature. The forming voltage must be considerably higher than the proposed operating voltage, to provide adequate dielectric strength over a long operating life despite aging effects. Leakage current increases rapidly as the operating voltage approaches the forming voltage value, particularly in "wet" electrolytics.

Needless to say, the foil, the electrolyte, and the tanks and apparatus must all be of the highest purity and cleanliness. The presence of impurities can result in porosity in the oxide film, and can cause some dissolving of the film in the electrolyte—an effect that can double for every $10°C$ rise in the temperature of the solution. Impurities remaining in the elements of a finished capacitor will cause reactions that will result in high leakage, early deterioration, and outright failure after a short operating life.

The Electrolyte

In an electrolytic capacitor, the electrolyte forms the second electrode, or plate. It is separated from the anode, or positive plate by the barrier layer of oxide formed on the anode surface. Ideally, it must be chemically inert, and have good temperature stability, and the proper conductivity. If the conductivity is too low, a high ESR (equivalent series resistance) results, with consequent high loss factor. If the conductivity is too high for the rated operating voltage, electrolytic breakdown in the form of sparking occurs (known as "scintillation"), resulting in failure of the capacitor.

"Wet" electrolytic capacitors use a liquid electrolyte, composed of a solvent (usually from the glycol family), some form of conductive salts, and a controlled amount of water. A porous ribbon of a nonconductive material such as a highly absorbent paper is wound as a separator between the two foils, and this ribbon is saturated with the electrolyte. The construction of such a capacitor is shown in Fig. 4.2. The rolled element is installed in a cylindrical metal container which may be connected to the cathode foil. A plastic sleeve is provided on some types, to facilitate the use of off-ground applications.

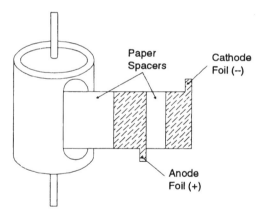

Fig. 4.2 Basic Construction of Electrolytic Capacitor

Limiting the amount of water in the electrolyte limits gassing and chemical activity, thereby increasing life expectancy. Low water levels in electrolytes also increase the shelf life. Using solvents less viscous than glycols, as an example amides, and more soluble salts, enables the electrolyte to penetrate into the fine etch structure of the foil more readily, thus contacting a greater surface area. This increases the ratio of unit capacitance to contact area and further reduces ESR.

Low viscosity electrolytes have additional desirable characteristics; they maintain more uniform conductivity at higher frequencies thus enhancing their utility as components in switching regulated power supplies. The lower ESR also enables these units to handle higher ripple currents than previously permitted.

The resistivity of electrolytes, particularly with high water content, varies with temperature especially below +25°C. This results in high loss of capacitance and increase in ESR at temperatures below −10°C. The high water content in electrolytes limits their use to a maximum temperature of +85°C. At these high temperatures they have limited life expectancy and shelf life. Local sites in the aluminum foil are activated by water, allowing exposure of bare metal. This results in high leakage current when a stored capacitor is subjected to applied voltage. If this leakage current can reach a sufficient magnitude from an unregulated source, the unit may go into thermal runaway, with subsequent failure.

At extremely low temperatures, conventional glycol family electrolytics lose in excess of 35% of their capacitance, and increase their ESR many fold, in relation to their room temperature values. These effects are caused by the increase in resistivity of the electrolyte, due to increased viscosity or sometimes by crystallization, as well as its shrinkage from the etch pattern of the foil. This results in poor contact between the electrolyte and the foils. This type of capacitor becomes a practically pure resistive device at −60°C due to these effects.

While it is a fact that amide based electrolytes produce capacitors with superior low temperature characteristics, such as an 80% capacitance retention at −55°C, they have several undesirable characteristics. The vapor pressures of amide base electrolytes are much higher than for glycols, thereby requiring superior sealing characteristics, and special materials in their containers. The high vapor pressures affect the long term life at high temperature operation. The toxicity of the amides is also considerably greater than glycols and they may also have adverse ecological effects.

The Spacer

The characteristics of the spacer that separates the foil ribbons influences the ESR of the capacitor. Each type of spacer has a resistance factor dependent upon its density, type of fiber, and fiber shape. In the design of low ESR capacitors, it is essential to use spacers with low resistance factors. At present, the lowest resistance factor spacers are the lowest density types, and since this low density is associated with minimum mechanical strength, special equipment is required to utilize them effectively. Proper design sometimes involves use of more than one spacer for optimum electrical characteristics and provides improved manufacturability.

The Cathode

The cathode foil in an electrolytic capacitor serves as a means of making extended contact with the electrolyte throughout the length and breadth of the separator strip. However, it also effectively forms an additional capacitor with the electrolyte, in series with the anode capacitor. The total effective capacitance is:

$$\frac{1}{C\ \text{total}} = \frac{1}{C\ \text{anode}} + \frac{1}{C\ \text{cathode}}$$

Theoretically, the cathode foil has no insulation or oxide coating; its capacitance therefore should be infinitely large, and total capacitance would be governed by the anode alone. Actually, a thin oxide film of some sort forms on the metal through exposure to the atmosphere and to the electrolyte, reducing this capacitance. In all but low voltage electrolytics, the cathode capacitance is considerably higher than the anode capacitance because of the relative thinness of the cathode film.

Nonpolar electrolytic capacitors are essentially two capacitors which are connected back to back. Both foils are anodized to form oxide barrier layers, and they share a common electrolyte. The system is inefficient, having a high power factor due to the large ESR, but is an effective and economical device in such AC applications as motor-start capacitors where intermittent use allows time for dissipation of heat between operations. The capacitor manufacturer is well equipped to advise a customer on the specification and application, to maximize performance at minimum cost, and to adjust the process to produce the idealized capacitor.

Electro-mechanical Considerations

Several mechanical innovations have been developed in electrolytic capacitor design by improving the electrical performance as well as the efficiency of these components. A very significant one is the multi-tabbing of the foil windings. Since the foil cross-section is very small, the foil resistance can be appreciable, especially in the larger diameter units. An effective method for minimizing this resistance is to install several connection tabs at equal distances along the length of the foils. This has the effect of connecting the resistance of the segments in parallel, thereby reducing the total resistance of the foil ribbon, and lowering the ESR.

The coiled helices of foil, being effectively in series with the conductive paths in the capacitor, also contribute some inductance to the ESL (equivalent series inductance) of the unit. Multi-tabbing reduces this effect significantly, not only by connecting the inductances of the segments in parallel, but also by the bifilar action of the centered tabs. For most effectively minimizing ESR and ESL, the tabs must be placed in the exact mathematical center of each segment; this placement is now accomplished by computerized techniques, which locate the tabs for optimum electrical performance, and for mechanical ease of assembly. Fig. 4.3 shows this multi-tabbed construction. Capacitors utilizing this construction can attain ESR values of milliohms in the 120Hz - 40KHz frequency range.

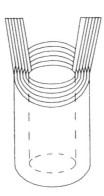

Fig. 4.3 Multi-Tabbed Doughnut-Wound Capacitor

Another benefit of the multi-tabbing techniques is greater realization of capacitance in high capacitance, low voltage units. Since the unit capacitance with its associated individual foil resistance is a strip-line network, reduction of the resistance increases the effective capacitance at the terminals of the device.

Another recent improvement is the elimination of potting compound. This is available on special order when a higher than standard vibration is a requirement. In previous designs, a bituminous compound was used to anchor the capacitor element in its metal case to prevent damage or failure due to mechanical vibration; this compound, which is a poor thermal conductor, constituted a barrier to efficient heat dissipation from the element.

Direct and firm contact between case and element provides excellent thermal conductivity from the element to the ambient atmosphere and chassis or frame support, resulting in cooler operation of the capacitor. In addition, absence of potting material results in uniform gas expansion space in each unit, increasing the operating life of the capacitor. Units with this construction are the most suitable for mounting in any plane during operation.

Further enhancement of the thermal efficiency of the unit is achieved by winding the capacitor element with a large core opening as shown in the Fig. 4.3. The development of foils with higher capacitance per unit area decreases the foil length required for a given capacitance rating. By keeping the outer diameter constant, for efficient roll contact with the case, and increasing the internal core size, a higher thermal efficiency is realized.

Thermal efficiency can also be boosted by judicious use of forced-air cooling in some applications. Occasionally, a power supply designer will calculate the proper capacitance value for the filter, then find that the ripple current requirements exceed the capability of the selected filter capacitor. The need for adding more capacitance in this situation can be avoided if forced air is used to increase the ripple current capability of the capacitor.

As shown in Fig. 4.4, the thermal resistance from the case of the capacitor to ambient can be reduced to less than 30% of the still air value. Another factor which is also shown is that the optimum air velocity is from 300 to 400 feet per minute. Air flow greater than 400 feet per minute causes very little decrease in thermal resistance.

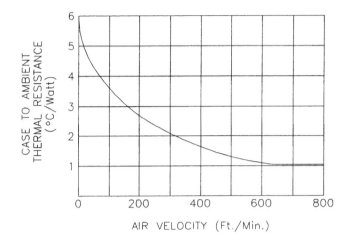

Fig. 4.4 Thermal Resistance vs. Air Velocity

The case to ambient thermal resistance in °C/Watt can be approximated by the following equation:

$$\theta ca = \frac{K}{A}$$

Where: θca = Thermal resistance case to ambient (°C/Watt)
K = 165 for still air
K = 64 for forced air at 300 ft./min. velocity
A = Surface area of metal case (in^2)

The graph in Fig. 4.5 demonstrates how more power can be dissipated in the components when forced air is used. The ratio of the two curves is 2.6, which means that 2.6 times the still air power can be dissipated with 300 ft./min. forced air. In aluminum electrolytic capacitors this gives a 50% increase in ripple current capability.

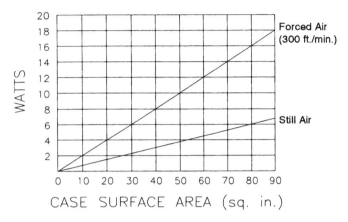

Fig. 4.5 Power Dissipation vs. Case Surface Area

Useful life expectancy is a function of the rate of electrolyte loss by means of vapor transmission through the end seal, and the rate of loss is a function of the operating or storage temperature. The electrolyte loss rate is relatively insensitive to operating voltage, provided the operating voltage does not exceed the rated voltage. Electrolyte loss is not related to the failure rate during the useful life period.

The typical reliability life cycle for aluminum electrolytic capacitors is shown in Fig. 4.6.

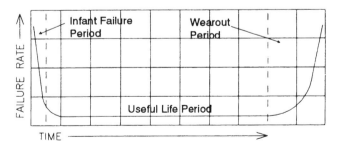

Fig. 4.6 Reliability Life Cycle - Typical Bathtub Curve

The plot of failure rate follows a characteristic "bathtub" curve, covering three periods in the typical capacitor life cycle.

The first period is the "Infant Failure" period, showing a decreasing failure rate. The manufacturer conditions and screens all capacitors to prevent or remove failures during this period. For all practical purposes, infant mortality is not a factor in shipped units.

The second segment in the life cycle is the "Useful Life" period. Failures occur at a constant rate on a random basis with relatively low frequency. Failure modes are related to temperature and voltage stress on the dielectric oxide film. Electrical parameters of capacitors are relatively constant during this period.

Electrolyte vapor transmission through the end seal occurs continuously throughout the useful life of the capacitor. This electrolyte loss has no effect on reliability during the useful life period of the life cycle, but when the electrolyte loss approaches 40% of the initial electrolyte content of the capacitor, the electrical parameters deteriorate and the capacitor is considered to have worn out.

The final period in the reliability life cycle is the "Wearout" period, exhibiting a rapidly increasing failure rate.

When aluminum electrolytic capacitors were first developed, "on-the-shelf" deterioration was a major problem and frequent replacement of stock parts was necessary. Additionally, use of capacitors for extended periods at small percentages of rated voltage permitted the dielectric oxide film to "deform," just as it would "on-the-shelf." Both problems were solved in the early 1950's with the introduction of high purity aluminum foil. Oxide film stability was greatly enhanced and today aluminum electrolytics can be used after storage, and at any percentage of rated voltage, without loss of capacitor quality. However, electrolyte loss through the end seal does occur during shelf storage and during periods of low voltage stress in operation, with predictable effects on the useful life of aluminum electrolytic capacitors.

Application Information

Aluminum electrolytic capacitors are intended for use in filter, coupling, and bypass applications where large capacitance values are required in small cases and where excesses of capacitance over the nominal value can be tolerated.

Nonpolarized capacitors should be used in applications where reversal of potential occurs.

Polarized capacitors should be used only in DC circuits with polarity properly observed. If AC components are present, the sum of the peak AC plus the applied DC voltage should never exceed the DC rating. The peak AC value should also be less than the applied DC voltage so that polarity may be maintained, even on negative peaks, to avoid overheating and damage. Capacitors which have been subjected to voltage reversal should be discarded.

Aluminum electrolytic capacitors provide the smallest volume, mass, and cost per microfarad of any type of capacitor with the exception of the tantalum electrolytic capacitor. These capacitors are nonhermetically sealed but do use elastomer seals. As is common with all electrochemical devices containing water, these capacitors can evolve small amounts of hydrogen during operation which, in most cases, is not sufficient to be considered hazardous.

It is recommended that capacitors having more than three years storage be checked at room temperature for leakage current in accordance with the application requirement before being placed in service.

Aluminum electrolytic capacitors are not suitable for airborne equipment applications since they should not be subjected to low barometric pressure and low temperatures at high altitudes. These aluminum electrolytic capacitors can be derated only for a short period since derating for any length of time may result in the necessity for re-forming. Even though they have vents designed to open at dangerous pressures, explosions can occur because of gas pressure of a spark ignition of free oxygen and hydrogen liberated at the electrodes. Provisions should be made to protect surrounding parts.

Aluminum electrolytic capacitors are generally used where pulsating, low frequency, DC signal components are to be filtered out. A typical application would be in vacuum tube circuit B power supplies, up to 350V or more DC working volts, at such points as plate and screen connections to B+, and as cathode bypass capacitors in self-biasing circuits. These capacitors are designed for applications where accuracy of capacitance is relatively unimportant.

As a rule, for selection of cathode or emitter by-pass capacitors, a ratio of bias resistance to by-pass reactance of about 10 to 1 is allowed. Ratios up to 20 to 1 may be used in high fidelity amplifier work or where space and economical considerations permit.

Electrolytic capacitors provide the equipment designer with unusually lightweight components of high capacitance in a compact container.

The most common failure mode of aluminum electrolytic capacitor types is typically a gradual loss of capacity and increased dissipation factor.

The 4-terminal axial leaded capacitor is designed for high frequency applications where low equivalent series resistance, inductance, and impedance are required. The advantage of the 4-terminal construction over 2-terminal construction is that the impedance decreases above 10 KHz. Unlike the 2-terminal capacitors, the DC current flows through the capacitor and contributes to the operating temperature. The ability of the external leads to carry the desired current should be taken into consideration. Lead length and heat sink qualities of the printed circuit board and capacitor will affect the current capability.

The thickness of the oxide film which is formed both initially on the foil and during the forming operations on the completed capacitor determines the maximum peak or surge voltage which may be applied.

For maximum reliability and long life, the DC working voltage should not be more than approximately 80 percent of full rating so that surges can be kept within the full-rated working voltage. The time of surge voltage application should not be more than 30 seconds every 10 minutes.

The metal cases for these capacitors are provided with an insulating sleeve. It should be noted that the insulating resistance refers to the sleeve and not to the resistance between the terminals and the case.

Even though these capacitors have vents designed to open at dangerous pressures, explosions can occur because of gas pressure or a spark ignition of free oxygen and hydrogen liberated at the electrodes. Safety provisions should be made to protect surrounding parts.

The surge voltage is the maximum voltage to which the capacitor should be subjected under any condition. This includes transients and peak ripple at the highest line voltage.

Aluminum electrolytic capacitors have poor resistance-temperature characteristics. As the temperature is raised, the breakdown voltage decreases and the leakage current increases.

In planning the location of the capacitors with respect to other circuit components, careful consideration should be given to the proximity of the capacitors to transformers, electron tubes, and high power resistors, because of the usual temperature rise involved in these components. Continued operation at temperatures above the normal rating will cause a permanent decrease in capacitance and an increase in series resistance.

The performance of capacitors at subzero temperatures is primarily affected by an increase in series resistance and a capacitance decrease. These changes do not persist with the return of normal temperature conditions.

Reliability is a measure of the expected failure rate during the useful life of the capacitor, and is dependent on the operating temperature and voltage.

The cost of a capacitor with an extremely long useful life may not be justified if the end product has a short useful life. The degree of reliability, therefore, is predicated on the "planned life" of the end equipment.

Generally speaking, there are four identifiable levels of reliability, they are expressed in terms of required years of operating life:
- Commercial - Consumer/industrial applications with 3-5 years of life.
- General Purpose - Industrial/data processing applications with 5-10 years of life.
- Long Life - Established-reliability applications requiring 10-20 years of life (–40°C to +85°C).
- High Reliability - Established-reliability applications requiring 10-20 years of life (–55°C to +125°C).

If the capacitor is used with reverse polarity, the oxide film is forward biased and offers very little resistance. The resulting high current, if left unchecked, will cause overheating and self destruction of the capacitor.

Electrolytic capacitors may fail for a number of reasons. One of the main causes of failure is the eventual drying out of the electrolyte. This results in a decrease in capacitance, an increase in dissipation factor or, at worst, an open circuit.

Short circuits in electrolytic capacitors have become uncommon, since potential shorts are generally weeded out during the manufacturing process.

Catastrophic failures like open or short circuits generally will render equipment totally inoperable or ineffective.

The true end of life for an electrolytic capacitor depends on its use. In one application a change of 10% in capacitance may be unacceptable, whereas in another circuit even a change of 25% may be tolerated.

Computer-grade aluminum electrolytic capacitors are designed for circuits requiring high CV, wide operating temperature range, and low ESR. Applications include filtering, data-system and industrial controls, and switch-mode power supplies.

Miniature aluminum electrolytic capacitors are suitable for use in timing and delay circuits as well as filtering, coupling, and decoupling. Typical applications include switching power supplies, consumer, industrial, data processing, and military usages.

Miniature aluminum electrolytic capacitors, with their extremely low leakage current, make them suitable replacements for many circuit applications using tantalum capacitors.

Electrolytic capacitors for DC applications require polarization. When installed, correct polarity should be confirmed. If used in reversed polarity, the circuit life may be shortened or the capacitor may be damaged. For use in circuits whose polarity is occasionally reversed, or whose polarity is unknown, use bi-polar capacitors.

If a voltage exceeding the capacitor's voltage rating is applied, the capacitor may be damaged as leakage current increase. When using the capacitor with AC voltage superimposed on DC voltage, care must be exercised that the peak value of AC voltage does not exceed the rated voltage.

Use the electrolytic capacitor at current values within the permissible ripple range. If the ripple current exceeds the specified value, request capacitors for high ripple current applications from the manufacturer.

Use electrolytic capacitors according to the specified operating temperature range. Usage at lower temperature will ensure longer life.

The standard electrolytic capacitor is not suitable for circuits in which charge and discharge are frequently repeated. This may cause the capacitance value to drop or damage the capacitor.

If the electrolytic capacitor is allowed to stand for a long time, its working voltage is liable to drop, resulting in increased leakage current. If the rated voltage is applied to such a product, a large leakage current occurs and this generates internal heat, which damages the capacitor. If the electrolytic capacitor is allowed to stand for more than 2-3 years, apply a voltage treatment before use. (Note: The voltage treatment is to gradually increase the applied voltage up to the rated value so that the current is set to the capacitor's specified leakage current value, and then keep the applied voltage at the rated value for 30 to 60 minutes.)

Some solvents have adverse effects on capacitors. An inquiry to the manufacturer is recommended to identify suitable cleaning agents.

The electrolytic capacitor is covered with a vinyl sleeve to insulate the case. If a soldering iron comes in contact with the electrolytic capacitor body during wiring, damage to the vinyl sleeve and/or case may result in defective insulation, or improper protection of the capacitor elements.

When soldering a printed circuit board with various components, care must be taken that the soldering temperature is not too high and that the dipping time is not too long. Otherwise, there will be adverse effects on the electrical characteristics and insulation sleeve of electrolytic capacitors. In the case of small-sized electrolytic capacitors, nothing abnormal will occur if soldering is performed at less than +260°C for less than 10 seconds.

If excessive force is applied to the lead wires and terminals, they may be broken or their connections with the internal elements may be affected.

The life of electrolytic capacitors may be adversely affected if exposed to high temperatures caused by such things as direct sunlight when in a storage area. Storage in a high humidity atmosphere may affect the solderability of lead wire and terminals.

The surge voltage rating is the maximum DC overvoltage to which the electrolytic capacitor may be subjected for short periods of time. In typical electrolytic capacitors, the applied voltage may not exceed the surge rating for more than 30 seconds and at infrequent intervals of not more than five minutes.

AC motor-start capacitors are often nonpolar aluminum electrolytic capacitors designed for intermittent AC duty; more specifically, the starting of small AC motors. They are not suitable for most DC or continuous AC applications.

The duty cycle of an AC motor start capacitor may be determined by dividing the capacitor's on-time (energized time) by the sum of its on-time (energized time and its off-time (de-energized time). For a given AC motor start capacitor, operating at a given voltage and ambient temperature; the time-averaged power dissipated by the capacitor, the internal operating temperature of the capacitor and, therefore, the life expectancy of the capacitor are all directly proportional to the capacitor's duty cycle.

Normal capacitor life may be realized (assuming voltage and temperature limits are not exceeded) when the on-time of a capacitor does not exceed 3 seconds and its duty cycle does not exceed 0.0167. Example: Twenty (20), three (3) second starts per hour yields a duty cycle of 0.0167 and does not exceed the three (3) second on-time limit.

Longer than 3 second on-times are not recommended as they will cause the capacitor's life to be shortened. Should they be unavoidable, there are certain things that can be done to minimize the degradation of the capacitor's life expectancy. For on-times up to 6 seconds:
- Reduce the duty cycle by increasing the off-time
- Reduce the ambient temperature
- Provide forced air cooling
- Use a capacitor with a higher voltage rating
- Series two capacitors each having twice the MFD value of the original.

AC motor start capacitors are designed for and are tested at 60Hz. They are, however, suitable for use from 50 to 60Hz.

The rated voltage is the rms value of AC voltage at which the capacitor may be operated at its normal duty cycle and maximum ambient temperature. During the start cycle of a normal capacitor motor, the voltage impressed across the AC motor start capacitor does not remain constant. It should start close to the rated voltage, dip slightly and then begin to increase as the motor's RPM increases. Should the start switch fail to open, it is possible for the capacitor's voltage to increase to as much as 2 to 3 times the capacitor's rated voltage.

As the temperature decreases from room temperature, capacitance starts to fall and % power factor (measurement of losses) starts to rise. Either one of these effects will cause a decrease in a motor's starting torque. The effects are such that below -40°C, a stalled rotor condition could occur. However, because the losses are so high, the internal capacitor temperature will rise rather quickly, thus restoring normal start torque. The total effect may just be a delay in the motor reaching switch-out speed.

The watt-second capability of AC motor start capacitors is high enough that precautions should be taken during the testing and application of these capacitors. Normally, the DC series resistance of the main and auxiliary windings are such that the capacitor is completely discharged prior to the motor coming to a complete stop. However, if this is not the case, or if this is deemed inadequate, discharge resistors should be used.

Some specialized applications require that the motor start capacitor be discharged prior to the closing of the start switch. This minimizes shock hazard, switch bounce noise, and peak contact currents. The resistor used to discharge the capacitor should be large enough so as not to significantly increase the power factor and small enough to insure discharging the capacitor within the time required. Normally, a 15K ohm, ±20%, 2-watt resistor is used.

Vertical mounting of the AC motor start capacitor with the terminals up is recommended; however, horizontal mounting with the pressure relief vent up is acceptable. Vertical mounting with terminals down or horizontal mounting with a relief vent down is not recommended as they may reduce capacitor life and could impair the operation of the pressure relief vent.

Recommended cleaning solvents are those free of halogens or halogen groups. Examples such as ethyl alcohol, butyl alcohol, methyl alcohol, propyl alcohol and deionized or distilled water are acceptable.

Solvents that are NOT recommended are halogenated hydrocarbon solvents such as Freon TF®, Freon TMC®, carbon tetrachloride, chloroform, trichloroethylene, trichloroethane, and methylene chloride. (®Registered Trademark EI DuPont & Company.)

WARNING: Misapplication, such as exceeding design limits or applying continuous AC voltage, may result in destruction or explosion of capacitors.

SELECTION CRITERIA FOR AC MOTOR START CAPACITORS

•Capacitor information: MFD value, voltage, etc.	•Any special or unusual application characteristic
•Case size (diameter x length)	•Maximum switch-out voltage
•Capacitors worst case duty cycle	•Maximum ambient temperature

Chapter 5

Tantalum Capacitors

Tantalum electrolytics have become the preferred type where high reliability and long service life are paramount considerations.

Tantalum is not found in its pure state. Rather, it is commonly found in a number of oxide minerals, often in combination with Columbium ore. This combination is known as "tantalite" when its contents are more than one-half tantalum. Africa and Brazil are the largest sources of tantalum, providing roughly 40% of the world's requirements. Canada, Australia, Malaysia, Thailand, and Mozambique also produce large quantities of this extraordinary metal.

After tantalum is mined, it goes through complex extraction processing which, at elevated temperatures, reduce the oxide to tantalum powder. Powder metallurgy techniques are used to make plates, sheets, and wire for the manufacture of chemical processing equipment, aerospace equipment, nuclear reactor parts, and elements of electron tubes.

Manufacturers use tantalum wire and foil, both of which are made from the powder, and also use pressed and sintered slugs of the powder itself as capacitor elements.

Tantalum capacitors contain either liquid or solid electrolytes. The liquid electrolyte in wet-slug and foil capacitors—generally sulfuric acid—forms the cathode (negative) plate. In solid electrolyte capacitors, a dry material, manganese dioxide, forms the cathode plate.

The anode lead wire from the tantalum pellet consists of two pieces. A tantalum lead embedded in, or welded to, the pellet, which is welded, in turn, to a nickel lead. In hermetically sealed types, the nickel lead is terminated to a tubular eyelet. An external lead of nickel or solder-coated nickel is soldered or welded to the eyelet. In encapsulated or plastic encased designs, the nickel lead, which is welded to the basic tantalum lead, extends through the external epoxy resin coating or the epoxy end fill in the plastic outer shell.

Tantalum Foil Style

Tantalum foil capacitors consist of a tantalum foil, acting as the anode, which is electromechanically treated to form a layer of tantalum oxide dielectric. Porous spacer material is used to form a cylindrical capacitor section with axial tantalum wire on either end. This section is then impregnated with a suitable electrolyte (usually a weak acid or base) and then sealed in a suitable container. Solderable leads are welded to the tantalum leads.

Tantalum foil capacitors are made by rolling two strips of thin foil, separated by a paper saturated with electrolyte, into a convolute roll. The tantalum foil which is to be the anode is chemically etched to increase its effective surface area, providing more capacitance in a given volume. This is followed by anodization in a chemical solution under direct voltage. This produces the dielectric tantalum pentoxide film on the foil surface.

Tantalum foil capacitors can be manufactured in DC working values up to 300 or more volts. The tantalum foil design has the lowest capacitance per unit volume of the three types of tantalum electrolytic capacitors. It is also the least often encountered since it is best suited for the higher voltages primarily found in older designs of equipment and requires more manufacturing operations than do the two other types. However, it is more expensive and is used only where neither a solid electrolyte nor a wet-slug tantalum capacitor can be employed.

Tantalum foil capacitors are generally designed for operation over the temperature range of $-55°C$ to $+125°C$ and are found primarily in industrial and military electronics equipment.

Wet Tantalum Style

Wet tantalum capacitors consist of a sintered slug, acting as the anode, which is electrochemically treated to form a layer of tantalum oxide dielectric.

Wet electrolyte, sintered anode tantalum capacitors, often called "wet-slug" tantalum capacitors, use a pellet of sintered tantalum powder to which a lead has been attached. This anode has an enormous surface area for its size because of the way it is made.

For wet-slug capacitors the tantalum powder of suitable fineness, sometimes mixed with binding agents, is machine pressed into pellets. The next step is a sintering operation in which binders, impurities, and contaminants are vaporized and the tantalum particles are sintered into a porous mass with a very large internal surface area. Following the sintering and before formation of the dielectric film on the pellet, a tantalum lead wire is attached by welding the wire to the pellet. (In some cases, the lead is embedded during pressing of the pellet before sintering.)

A film of tantalum pentoxide is electrochemically formed on the surface areas of the fused tantalum particles. Provided sufficient time and current is available, the oxide will grow to a thickness determined by the applied voltage.

The pellet is then inserted into a tantalum or silver can which contains an electrolyte solution. Most liquid electrolytes are gelled to prevent the free movement of the solution inside the container and to keep the electrolyte in intimate contact with the capacitor cathode. A suitable end seal arrangement prevents the loss of the electrolyte.

Wet-slug tantalum capacitors are manufactured in a voltage range up to 150 Working Volts DC.

Solid Tantalum Style

The solid tantalum capacitor is generally included in the classification of "electrolytic" capacitors, although it doesn't belong there. A true electrolytic capacitor is one which uses an electrolyte for at least one of the electrodes. An electrolyte is generally made from some chemically ionizable compound dissolved in a liquid. The solid tantalum (also sometimes called a dry tantalum capacitor) uses manganese dioxide, rather than a liquid electrolyte as an electrode. Because of similar characteristics and because of historical development, the manganese dioxide came to be a solid electrolyte, but it really is not; it falls generally into the class of semiconducting solids.

The basis of the solid tantalum capacitor is tantalum. Tantalum is one of the elements, so named because it was a tantalizing material for early chemists to isolate. It has certain properties which produce the characteristics found so desirable in finished capacitors. Foremost, is the fact that it is a "valve" metal (aluminum is another) upon which one may grow very uniform and stable oxides with good dielectric properties. The dielectric constant of tantalum oxide ($K = 26$) is relatively high. To form a high quality oxide film requires very high purity of the metal substrate. Tantalum melts at +3000°C and can be worked above +2000°C in vacuum. Under these conditions, most of the impurities can be evaporated and pumped away. Finally, tantalum is relatively easy to work with mechanically. That is, it can be ground to powder, rolled to sheet, drawn to wire, bent and formed without great difficulty at room temperature.

To make capacitors from this material, manufacturers use tantalum powder and tantalum wire. These two are pressed together, usually with some form of organic binder which is later removed. The pressed form is normally cylindrical, although flat prisms appear in some designs.

While it is also possible to press only the powder and weld on the wire later, the cylindrical wire-and-powder assembly shown in Fig 5.1 is by far the most popular method. These pellets are sintered in vacuum furnaces. Sintering is a process of slow fusion between adjacent surfaces, so that when the pellets emerge from the furnace they are strong mechanically and have shrunk somewhat from their original size. About half the volume of the sintered powder remains as void space.

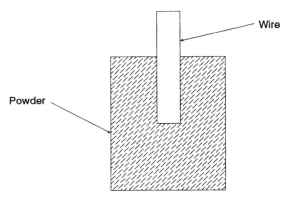

Fig. 5.1 Pressed and Sintered Tantalum

The pellets are then immersed in an acid bath and connected to the positive terminal of a DC power supply as shown in Fig. 5.2.

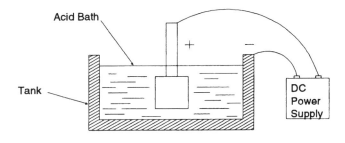

Fig. 5.2 Tantalum Pellet Immersed in Acid Bath

The flow of current causes a layer of tantalum pentoxide, Ta_2O_5, to grow on all exposed surfaces of the tantalum. The exposed surface includes the wire and both external and internal surfaces of the sintered powder. The internal surface is over 100 times the apparent external area. The oxide layer later will become the dielectric of the capacitor. One of the electrodes of the parallel-plate model is the tantalum metal; the second electrode will be applied in subsequent processing steps. The effective area of the capacitor is the entire surface of the tantalum pentoxide dielectric which can be contacted by the second electrode. The distance between electrodes is the thickness of the oxide layer. This thickness is controlled by the voltage applied from the power supply. The higher the voltage, the thicker the oxide layer grows. Greater separation between electrodes means lower capacitance, but it also means a higher voltage rating for the finished capacitor.

The second electrode is the semiconducting manganese dioxide, MnO_2. To apply this material, the porous pellet is dipped into a manganous nitrate solution which wets all surfaces and fills up the pores. When the pellets are later heated, the water from this solution is evaporated and then the nitrate decomposes to form the oxide according to this chemical equation:

$$Mn(NO_3)_2 \xrightarrow{heat} MnO_2 + 2NO_2 \text{ (Gas)}$$

The MnO_2 layer covers nearly all the internal surfaces and extends part way up the wire.

Fig. 5.3 illustrates a small portion of the pellet. Shown is the tantalum substrate, the tantalum pentoxide that is grown upon the substrate, and the manganese dioxide deposited upon the tantalum pentoxide. It then begins to look like a parallel-plate capacitor.

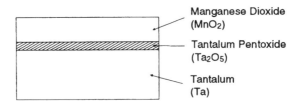

Fig. 5.3 Section of a Tantalum Pellet

The rest of the processing is needed only to gain electrical contact to the electrodes. It is easy to weld an external lead wire to the stub of the tantalum wire, but contacting the MnO_2 is more difficult. To do this, the pellets are dipped into water containing a very finely divided carbon powder. After the water is evaporated, a layer of carbon (actually graphite) is left on all surfaces of the MnO_2. Resistivity of the graphite is much lower than that of MnO_2, and the fine particle size of the graphite enables this material to touch nearly all the very irregular MnO_2 surface. Applied on top of the graphite is a silver-pigmented paint. The silver is held by an organic resin and presents a solderable surface to facilitate attachment of the second lead wire. Putting all the layers together looks like Fig. 5.4, with the two wires being indicative of external connection.

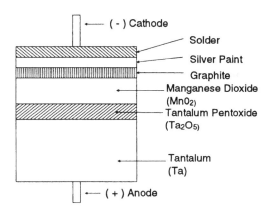

Fig. 5.4 Tantalum Dielectric Layers

Encapsulation of a solid tantalum capacitor can follow several courses. The original design was solder inside a metal can closed with a glass-to-metal hermetic seal. The next commercial design used potting with an epoxy, and then dipping in liquid epoxy resin. This offers excellent reliability and high stability for consumer and commercial electronic applications, with the added feature of low cost. The final step in evolution to this point is the tantalum chip capacitor, which has no encapsulation but which has several innovations in terminal design to provide protection against the rigors of directly soldering onto ceramic substrates.

Manufacturers have done much work into statistical treatment of failure rates of solid tantalum capacitors because these capacitors possess a unique "healing" mechanism which results in a failure rate apparently decreasing forever. The MnO₂ provides the healing mechanism. If a fault, perhaps some impurity produces an imperfection in the dielectric layer, a heavy current will flow through that minute area when a DC potential is applied to the capacitor. This is illustrated in Fig. 5.5.

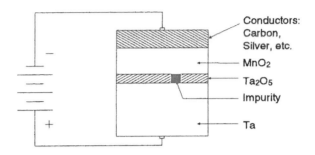

Fig. 5.5 Fault in a Tantalum Capacitor

The current also flows through the MnO₂ immediately adjacent to the fault. Resistance of the MnO₂ to this current flow causes localized heating. As the temperature of the MnO₂ rises, this material is then converted to a lower oxide of manganese, perhaps Mn₂O₃, with much higher resistivity. The increase in resistance decreases the current flow. If this mechanism is successful, the current flow is reduced before localized heating goes too far and a short circuit is prevented. Without this mechanism, the solid tantalum capacitor would never have gotten off the ground commercially.

Application Information

In choosing between the three basic types of tantalum capacitors, the circuit designer customarily uses foil tantalum capacitors only where high voltage constructions are required or where there is substantial reverse voltage applied to a capacitor during circuit operation.

Wet sintered anode capacitors, or wet-slug tantalum capacitors, as they are sometimes called, are used where the lowest DC leakage is required. The conventional silver can design will not tolerate any reverse voltages. However, in military or aerospace applications, tantalum cases are used instead of silver cases where utmost reliability is desired. The tantalum cased wet-slug units will withstand reverse voltages up to 3 Volts, will operate under higher ripple currents, and can be used at temperatures up to +392°F (+200°C).

Solid electrolyte designs, which are the least expensive for a given rating, are used in many applications where their very small size for a given unit of capacitance is of importance. They will typically withstand up to about 15% of the rated DC working voltage in a reverse direction. Also important are their good low temperature performance characteristics and freedom from corrosive electrolytes.

Tantalum capacitors may utilize only 15 percent of the area normally required by an aluminum/paper capacitor of the same capacitance value.

The larger the dielectric constant the larger the capacitance which can be achieved in a given space, thus a size advantage can be realized since the dielectric constant of tantalum oxide film is approximately 26 as compared to approximately 8 for aluminum oxide. Because of the differences in foil and paper thickness requirements, the actual size ratio will vary with different capacitances and voltage ratings and may be much more than 2:1 in favor of the tantalum capacitor.

Tantalum Foil Capacitors

The tantalum foil types are the most versatile of all the electrolytic capacitors. They are available in plain or etched-foil and in polarized or nonpolarized construction, which makes them suitable for many circuit applications; however, the foil types are limited by their great variation of characteristics and design tolerances. Because of the very wide tolerances, they are not suitable for timing or precision circuits.

Etched-foil types, because of the difference in construction, have as much as 10 times the capacitance per unit area as the plain-foil types for a given size; therefore, the etched-foil type is generally the better choice between the two.

Some etched-foil style capacitors may exhibit capacitance change and dissipation factor changes when exposed to low DC bias levels (0 to 2.2 Volts DC). Care should be exercised when applications require these low voltage levels.

Plain-foil types, in some cases, are a more desirable choice since they will withstand approximately 30 percent higher ripple current. They have better capacitance-temperature characteristics, and have a low power factor.

Tantalum foil capacitors are the only electrolytic capacitors capable of operating continuously on unbiased AC voltages. Tantalum foil's AC ripple capabilities are applicable for unbiased AC voltages on nonpolar capacitors, and biased AC ripple voltages on polar capacitors. Peak AC voltages are permissible provided that the DC voltage rating is not exceeded. The only limitation is the I^2R heating effect. Due to a higher power factor, etched-foil capacitors have only half the AC capability of plain-foil capacitors.

Polarized foil types are essentially used where low frequency pulsating DC components are to be bypassed or filtered out and for other uses in electronic equipment where large capacitance values are required and comparatively wide capacitance tolerances can be tolerated.

When polarized foil types are used for low frequency coupling in vacuum tube and transistor circuits, design allowances should be made for the leakage current. This leakage current could cause improper positive bias to be applied across the grid circuits or excessive base, emitter, or collector currents.

Polarized foil capacitors should be used only in DC circuits with polarity properly observed. If AC components are present, the sum of the peak AC voltage plus the applied DC voltage must not exceed the DC voltage rating. The peak AC voltage should also be less than the applied DC voltage so that polarity may be maintained, even on negative peaks, to avoid overheating and damage. Even though those units rated at 6 Volts and above can withstand a maximum of 3 Volts in the reverse direction, it is recommended that they not be used in circuits where this reversal is repetitious.

Polarized foil capacitors may be used:
- in power supplies in which up to 300 Volts DC are applied to the filter input
- at plate and screen circuit decoupling connection points
- for cathode resistor bypass circuits.

When polarized foil capacitors are used as cathode bypass capacitors, a ratio of bias resistance to capacitive reactance of 10 to 1 is allowed. Ratios up to 20 to 1 may be used in high fidelity amplifier design or where space and economic considerations permit. In circuits where linear amplification is required, the amount of capacitive reactance shunting a cathode resistor will depend on the percentage of feedback desired.

Nonpolarized foil capacitors have two polarized sections, their cathodes connected back to back, enclosed within an outer enclosure.

Nonpolarized foil capacitors are primarily suitable for AC applications or where DC voltage reversals occur. Examples of uses are:
- tuned low frequency circuits
- phasing of low voltage AC motors
- computer circuits where reversal of DC voltage occurs
- servo systems.

Wet-Electrolyte, Sintered Anode Tantalum Capacitors

Wet-slug capacitors are not suitable for applications involving any voltage reversal. They cannot be operated on unbiased AC voltage or applied in nonpolar applications involving back-to-back connections. Any AC ripple applied to wet slug capacitors must be superimposed on sufficient DC bias voltage to prevent voltage reversal. Ripple current is limited to small values because progressive degradation of the unit will result if the cathode (silver case) becomes positive during the discharge cycle.

Some wet-slug capacitors are for DC applications only — no reverse voltage can be tolerated. (The most common failure mode, electrolyte leakage due to seal failure, is primarily due to the application of reverse voltage.) Some styles can withstand a small amount of reverse voltage without damage, but care should be exercised when applications require these voltages.

Sintered wet-slug capacitors are limited to low voltage applications. Their primary use is in low voltage power supply filtering circuits. Their low leakage current (lowest of all the tantalum types) is not appreciable below +85°C. At typical operating temperatures they are comparable to good quality paper capacitors, yet are much smaller in size.

Hermetically sealed sintered-anode capacitors' primary applications are in industrial, military and aerospace equipment where reliability and premium performance with respect to low DC leakage current, high inrush current capability, and high volumetric efficiency are vital.

Tubular, elastomer-sealed, wet-slug (gelled-electrolyte) capacitors fill the basic requirements for applications where a superior quality and reliable design for industrial, automotive, and telecommunications applications is desired. These capacitors feature an extremely long shelf life in excess of ten years.

Tantalum-case hermetically sealed wet-slug capacitors are ideal for high-performance filter, bypass, coupling and timing applications for power supplies, computers, telecommunications, instrumentation and control systems. These capacitors are suited for use in applications where minimum size and weight conditions must be achieved and reverse voltage or high ripple currents are required.

The construction of tantalum-case hermetically sealed tantalum capacitors assures operational capability in deep oceanographic environments or space, where dependability is required for tens of thousands of storage or operating hours. To meet aerospace requirements, the capacitors have a high resistance to damage from shock and vibration.

Wet-slug capacitor assemblies are widely used in filter, coupling, bypass, and time-delay circuits in computers, missiles, airborne electronic equipment, radar, and fire control systems.

Whenever these tantalum capacitors are connected in series for higher voltage operation, a resistor should be paralleled across each unit. Unless a shunt resistor is used, the DC rated voltage can easily be exceeded on the capacitor in the series network with the lowest DC leakage current. To prevent capacitor destruction, a resistance value not exceeding a certain maximum should be used; this value will depend on capacitance, average DC leakage, and capacitor construction.

Foil or sintered slug tantalum capacitors may be connected in parallel to obtain a higher capacitance than can be obtained from a single capacitor. However, the sum of the peak ripple and the applied DC voltage should not exceed the DC rated voltage. The connecting leads of the parallel network should be large enough to carry the combined currents without reducing the effective capacitance due to series lead resistance.

Foil and sintered slug tantalum electrolytic capacitors have excellent life and shelf life characteristics. Life, at higher temperatures than with aluminum electrolytics, will show a comparatively lower decrease in capacitance. With rated voltage applied, more than 10,000 hours of life can be expected at +85°C. All styles may be expected to operate at least 2,000 hours at +85°C with less than 10 percent loss of capacitance.

Because the more stable tantalum oxide film is less subject to dissolving the surrounding electrolyte than the film in an aluminum capacitor, the shelf life of the tantalum unit is much longer, and less re-forming is required. After storage for long periods, the re-forming current is low and the time is comparatively short; it may be expected to take less than 10 minutes. These properties are affected by the storage temperature to a significant degree, being excellent at temperatures from −55°C to +25°C; good at +65°C; and relatively poor at +85°C.

The predominant mode of failure of the wet electrolytic tantalum capacitors will most probably be a "hidden" failure mode of high or erratic leakage current which could result in a catastrophic short circuit. This is a result of electrolyte migration past the inner seal and touching the cathodic case or ground.

When complex ripple wave shapes are involved, they should be measured on an oscilloscope or by some other method which will give the peak rating. Foil and slug capacitors should be limited to operation at ripple frequencies between 60 Hz and 10,000 Hz (above 10,000 Hz, the effective capacitance rapidly drops off). At frequencies of only a few hundred KHz, these tantalum units act as practically pure resistance.

Solid Tantalum Capacitors

Solid tantalum capacitors are intended for use in equipment where a known order of reliability is required. These capacitors are the most stable and reliable electrolytics available, having a longer life characteristic than any of the other electrolytic capacitors. Because of their passive electrolyte being solid and dry, these capacitors are not temperature sensitive; they have a lower capacitance-temperature characteristic than any of the other electrolytic capacitors.

Solid tantalum capacitors' limitations are the relatively high leakage current, limited voltage range available (6 to 100 Volts), and a maximum allowable reverse voltage of 15 percent of the rated DC voltage at +25°C to 1 percent at +125°C.

Solid tantalum capacitors are often used where pulsating low frequency DC components are to be bypassed or filtered out, such as output filtering for switching regulator power supplies. These uses require low impedance in series with the capacitors.

Solid tantalum capacitors are used in electronic equipment where large capacitance values are required, where space is at a premium, and where there are significant amounts of shock and vibration.

These capacitors are mainly designed for filter, bypass, energy storage, coupling, blocking, and other low voltage DC applications (such as transistor circuits in missile, aircraft electronics, and computers) where stability, size, weight, and shelf life are important factors.

When designing transistor, timing, phase shifting, and vacuum tube grid circuits, the dissipation factor and power factor should be taken into consideration.

For bypassing resistors, a ratio of bias resistance to capacitive reactance of 10 to 1 is usually allowed. Ratios up to 20 to 1 may be used in high fidelity amplifier work or where space and economic considerations permit. In circuits where linear amplification is required, the amount of capacitive reactance shunting a cathode resistor will depend on the percentage of degenerative feedback desired.

Polarized types should have their cases at the same potential as the negative lead; they should be used only in DC circuits with polarity observed.

Polarized types are especially adapted for applications in coupling, filtering, and timing in those computer, industrial, and commercial circuits requiring capacitors with more capacitance than other standard types.

In high impedance circuits, momentary breakdowns (if present) will self-heal; however, in low impedance circuits, their self-healing characteristics under momentary breakdown of the dielectric film will be nonexistent. The large currents in low impedance circuits will cause permanent damage to the capacitor.

Solid tantalum capacitors may be operated with an impressed ripple (AC) voltage provided the heat dissipation limits are not exceeded. Total heat dissipation limits are dependent on the ambient operating temperature and the operating frequency.

When solid tantalum capacitors are connected in series, the maximum voltage across the network should not be greater than the lowest voltage rating of any capacitor in the network. Voltage divider resistors could be used to prevent over voltage on one or more units of the series capacitor group.

Solid tantalum capacitors, when connected in parallel, will obtain a higher capacitance than can be obtained from a single capacitor. The sum of the peak ripple and the applied DC voltage should not exceed the DC working voltage of the unit with the lowest voltage rating. The connecting leads of the parallel network should be large enough to carry the combined currents without reducing the effective capacitance due to series lead resistance.

Dielectric absorption may be observed as the re-appearance of the potential across the capacitor after it has been shorted and the short removed. This characteristic is important in RC timing circuits, triggering systems, and phase-shift networks. The voltage recovery can be measured with a high-impedance electrometer.

Solid tantalum capacitors differ from aluminum electrolytics in several important aspects; namely, substantially indefinite shelf life, superior low temperature characteristics, complete freedom from electrolyte leakage, and higher operating temperatures. Solid tantalum capacitors generally are more costly than aluminum electrolytic capacitors. Some consideration should be given to the use of aluminum electrolytic capacitors if their performance characteristics and physical sizes are suitable and if the application will permit.

Failure rate is a function of temperature, applied voltage, and circuit impedance. Increased reliability may be obtained by derating operating temperature and applied voltage and increasing circuit impedances.

The DC leakage current increases when either voltage or temperature is increased; the rate of increase is greater at the higher values of voltage and temperature. A point can be reached where the DC leakage current will avalanche and attain proportions that will permanently damage the capacitor. Consequently, capacitors should never be operated above their rated temperature and rated voltage for that temperature.

By increasing the circuit impedance, the leakage current is reduced. In life testing the solid tantalum capacitor, the capacitance and dissipation factor are very stable over long periods of time and hence are not a suitable measure of deterioration. Leakage current variation is a better indicator of capacitor condition. It is recommended that a minimum circuit impedance of 1 ohm per applied volt be utilized to attain improved reliability.

Hermetically sealed solid tantalum capacitors in a metal case provide proven reliability in a wide variety of high-performance military, aerospace guidance and control, industrial, and commercial applications.

Molded case solid tantalum miniature capacitors provide the cost savings associated with nonmetal case technology. They are designed for use in automotive, industrial and commercial electronic equipment in circuits which include filtering, time constant, coupling, by-passing and energy storage applications.

Triple lead resin-coated solid tantalum capacitors make for simpler installation, economy, and high performance. The anode lead is the center; both outside leads are the cathode leads. This three lead design makes it impossible to insert the capacitor backwards, eliminating many of the problems of rework and board damage.

Nonpolarized design consists of two polar solid electrolyte sintered tantalum anode capacitors assembled into a tubular metal can in a back-to-back configuration.

Nonpolarized types should be used where reversal of potential occurs. These are especially suited for applications involving coupling, filtering, and timing in those computer, industrial, and commercial circuits where AC signals are prevalent, or where voltage reversals are common.

Nonpolarized capacitors feature miniature size, excellent electrical characteristics and outstanding service life.

Nonpolarized capacitors are used in many applications. Examples include:
- Phase splitting capacitors for small, low-voltage motors
- Servo systems
- Low frequency tuned circuits
- Crossover networks
- Bypass applications where high ripple voltages are encountered
- Circuits in which reversals of polarity are greater in magnitude than those which may be applied to a polarized device.

Tantalum chip capacitors are primarily intended for use in thick and thin film hybrid circuits for filter, bypass, coupling, and other applications where the alternating current (AC) component is small compared to the direct current (DC) rated voltage and where supplemental moisture protection is available.

In circuits where voltage reversals exceeding 15% at 25°C occur, or where AC ripple voltages are high, chips should be mounted in pairs in a cathode to cathode arrangement to effect nonpolar operation. In these cases, if two identical chips are mounted back to back, the resulting capacitance will be 1/2 the value of the original devices and the voltage rating will be the same as the rating of the original chips.

Chapter 6

Glass Capacitors

Glass dielectric capacitors offer the end user the highest performance and reliability features available in the capacitor industry.

The construction of glass capacitors is straightforward. There are only three elements: glass dielectric and case, aluminum electrodes, and wire terminals. The capacitors are made in a multilayer fashion with the leads being welded to the electrodes so there are no pressure connections to come loose and no solder connections to melt.

The dielectric is formed as a continuous ribbon of glass. Physical and electrical properties and dimensions are precisely controlled. This results in every glass capacitor being just like every other, part-to-part, and lot-to-lot. Couple this predictability with complete performance specifications and you know what performance to expect before the first prototype is built.

Glass dielectric capacitors are composed of alternate layers of glass ribbon and the electrode material. After assembly, the units are sealed together by high temperature and pressure to form a rugged monolithic block. Since the terminal leads are fused to the glass case, the seal cannot be broken without destroying the capacitor. Although these capacitors are of a monolithic structure, they are not necessarily hermetically sealed since the coefficient of thermal expansion of the terminals does not match that of the case.

Glass capacitors may be enclosed in glass or vitreous enamel cases, suitably protected against environmental conditions.

Glass capacitor construction features provide the following advantages:
- Fixed temperature coefficient
- High insulation resistance
- Low dielectric absorption
- Readily used where miniaturization is demanded
- Ability to operate in environments involving high humidity and high temperatures
- Extended life of 30,000 hours and more.

Capacitor stability, or lack of it, is an inherent characteristic of the dielectric used. Few materials can match glass for stability.

The reasons for the high reliability of glass capacitors are the simple construction with few things to go wrong, and the glass dielectric - one of the most stable, inert materials available.

Glass doesn't corrode or degrade in any way. Glass is not subject to microfractures, delaminations, and other problems associated with some crystalline materials. In addition, axial glass capacitors are hermetically encased in glass, with a true glass-to-metal seal at the leads. This construction is practically immune to severe environmental effects such as shock, vibration, acceleration, vacuum, radiation hardness, moisture, salt spray, and solder heat.

Glass capacitors are made of inorganic materials and are highly resistant to high operating temperatures, voltage breakdown, and nuclear radiation. Glass capacitors will not become a toxic hazard when exposed to radiation.

When exposed to gamma irradiation, glass capacitors exhibit a very small transient and permanent capacitance change and a consistently low dissipation factor with frequency.

Glass dielectric capacitors have been tested for Dielectric Absorption (DA) characteristics and have shown a consistently low Dielectric Absorption from lot to lot. The DA figures for glass compare favorably to polystyrene.

The combination of tight quality control, simplicity of design, and a superior material system means that glass capacitors are close to being the "perfect capacitor circuit symbol."

Glass capacitors are intended for use in any equipment where known orders of reliability are required, and are primarily designed as a substitute for mica dielectric capacitors as a step toward conservation of critical mica. They are effective substitutes for mica dielectric capacitors and can be employed for many applications where mica dielectric capacitors are used, provided consideration is given to the difference in temperature coefficient and dielectric loss.

Glass capacitors are capable of withstanding environmental conditions of shock, vibration, acceleration, extreme moisture, vacuum, extended life of 30,000 hours and more, and high operating temperatures such as experienced in missle-borne and space electronic equipment.

Here is an overview of glass capacitor features:
HIGH STABILITY
- Low Retraceable Temperature Coefficient (TC)
- No Hysteresis
- Zero Aging Rate
- Zero Piezoelectric Noise.

EXTREMELY LOW LOSSES
- Stable Q Factor at High Frequencies
- Low Dielectric Absorption.

LARGE RF CURRENT CAPABILITY
NUCLEAR RADIATION HARDNESS (Axials)
HIGH OPERATING TEMPERATURE RANGE (Up to 200°C)
HIGH SHOCK/VIBRATION CAPABILITY
EXCELLENT HIGH VACUUM PERFORMANCE

Application Information

Glass capacitors have traditionally seen widespread usage in military applications with a large number of new designs occurring in the aerospace and high performance commercial sectors. Glass capacitors have applications across the entire spectrum of electronic circuits and their past success on a variety of manned and unmanned space missions continues to fuel interest in the defense and aerospace industries.

In most types of electronic equipment, the occasional failure of a capacitor is tolerable - though it is inconvenient and often costly. For these applications, an acceptable level of reliability is provided by many of the excellent types of capacitors available today.

In a few designs, failures are not acceptable, such as satellite systems, undersea cable repeaters, and mountaintop microwave relay stations. For these designs, glass capacitors may be the optimum choice.

Where reliability is critical and replacement of a failed part is not possible or practical, consider glass capacitors.

Where stability is essential, even in severe environments, consider glass capacitors.

Glass capacitors have been used in virtually all critical military and space programs. Increasingly, they are also being used in commercial and industrial applications where failures can't be tolerated and circuit performance is critical.

Glass capacitors exhibit low loss over a wide operating temperature and frequency range.

Glass capacitors exhibit zero aging rate, zero piezoelectric noise, and a ± 5 ppm TC retraceability regardless of component age. Furthermore, glass capacitors exhibit zero voltage coefficient and low thermal and charge noise figures.

Glass capacitors continue to experience widespread usage in sample and hold current integrators, and in high gain amplifiers.

Glass capacitors can handle large Radio Frequency (RF) currents over a wide frequency range.

Glass dielectric capacitors have a high Q factor and a low dissipation factor that changes little with frequency and temperature excursions. This coupled with a low, retraceable, extended range temperature coefficient ensures repeatable, reliable performance - regardless of the capacitor's environment.

Glass capacitors are effective substitutes for mica dielectric capacitors, if consideration is given to the differences in temperature coefficient and dielectric loss.

Glass dielectric capacitors exhibit a much higher Q over a wider range of capacitance than mica dielectric capacitors.

The large RF currents that glass dielectric capacitors can handle make them ideal for use in modulators, filters, and linear amplifiers.

The glass capacitors with stable performance, coupled with an excellent frequency response and miniature size, make them a natural choice for oil well logging and downhole instrumentation systems, jet engine monitors, semiconductor burn-in ovens, or geophysical pressure probes.

Some other applications where glass capacitors have already been proven include:
- Missile or aerospace applications where engine or environmental heat needs to be monitored or may cause circuit failure.
- Radar or other microwave application.
- RF output circuitry where conduction or fan cooling cannot be entirely relied upon to remove all of the heat.
- Space and satellite applications where temperature changes are extreme and "zero failures" are a must.
- Industrial chemical process instrumentation where heat is a part of the process.
- Instrumentation for monitoring at-the-tool performance in metal cutting machinery.
- Fire-safe alarm or control circuitry.

Glass capacitors are resistant to high G loads but they are susceptible to damage from mild mechanical shocks and therefore should be mounted accordingly.

In general, glass capacitors are ideally suited for any environment where high temperature could alter or destroy circuit performance. Glass capacitors are also useful where cycling to colder temperatures may be a problem.

Chapter 7

Mica Capacitors

Muscovite mica is the most commonly used mica dielectric material. It has a dielectric constant between 6.5 and 8.5. Muscovite mica can be split into thin sheets; it is nonporous and does not readily absorb moisture. Protection from moisture is provided to obtain capacitance stability and low losses.

The two techniques used to form the capacitors are by stacking the mica sheets through the silvered-mica process or by the use of tin-lead foil to separate the mica sheets.

Terminals are attached to the mica stacks by the use of pressure clips which have been solder coated for maximum mechanical strength.

Button style mica capacitors are composed of a stack of silvered-mica sheets connected in parallel. This assembly is encased in a metal case with a high potential terminal connected through the center of the stack. The other terminal is formed by this metal case connected at all points around the outer edge of the electrodes.

The button style design permits the current to fan out in a 360° pattern from the center terminal. This provides the shortest RF current path between the center terminal and chassis. The internal inductance is thus kept small. The use of relatively heavy and short terminals results in minimum external inductance associated permanently with the capacitor. The units are then welded and hermetically sealed with either glass or resin.

The molded case is made of a polyester material which also exhibits high insulation resistance and high resistance to moisture absorption and transmission. The molded case also imparts rigidity to the capacitor in the event the capacitor is subjected to vibration or shock.

The DC voltage ratings are for continuous operation throughout the operating temperature range. At higher frequencies, the operating conditions are usually limited by the AC current rather than the voltage. Voltage ratings range from 300 to 2,500 volts. In addition to the limitations of operation placed on the capacitor by operating temperature range and AC current at high frequencies, the following conditions should also be considered:

BAROMETRIC PRESSURE:
- Up to and including 1,200 volts - 0.315 inches of mercury (100,000 feet) up to normal atmospheric pressure.
- Above 1,200 volts - 3.44 inches of mercury (50,000 feet) up to normal atmospheric pressure.

RELATIVE HUMIDITY - Up to 80 percent.

Application Information

The characteristics of mica dielectric are high insulation resistance and high breakdown voltage, low power factor, low inductance, and low dielectric absorption.

Mica capacitors are designed for the circuits requiring precise frequency filtering, bypassing, and coupling.

Mica capacitors are used where close impedance limits are essential with respect to temperature, frequency, and aging. Examples are in tuned circuits which control frequency, reactance, or phase.

Mica capacitors are also used as padders in tuned circuits, as secondary capacitance standards, and as fixed tuning capacitors at high frequency.

Mica capacitors can also be employed in delay lines and stable low power networks.

Due to the inherent characteristics of the dielectric, mica capacitors are inexpensive, small, and readily available and have good stability and high reliability.

Button style capacitors are intended for use at frequencies up to 500 MegaHertz (MHz). Their principal uses are in tuned circuits, and for the coupling and bypassing applications in VHF and UHF circuits.

Button style capacitors are very stable with time and have high reliability in circuits where ambient conditions can be closely controlled to reduce failure from silver ion migration. Due to this silver ion migration, silvered-mica capacitors should not be used under DC voltage stresses when combined with exposure to continuous high temperature and humidity conditions for extended periods.

Silver ion migration can occur in only a few hours when silvered-mica capacitors are simultaneously exposed to DC voltage stresses, humidity, and high temperatures.

Dipped silvered-mica capacitors are designed to meet the electrical requirements of the latest electronic equipment where size is critical. Applications are found in a diversity of high grade ground, airborne, and space borne devices, such as computers, jet aircraft and missiles.

Relatively low in cost, dipped mica capacitors are ideally suited to a variety of uses in many types of commercial and industrial apparatus where high stability characteristics, close capacitance tolerances, and high Q are prime factors. Such applications are tuned circuits, delay lines, and filter circuits in conventional or printed wiring assembles.

It is recommended that the capacitor have adequate heat sink during mounting operation with high temperature solder.

Glossary

AC - Alternating Current

Aging - Capacitor aging is a term to describe the negative, logarithmic capacitance change that takes place in ceramic capacitors with time. The more stable dielectrics have the lowest aging rates. Aging is also operating an aluminum electrolytic capacitor at controlled conditions of time and temperature to screen out weak or defective devices and at the same time stabilize the good units.

Aluminum Electrolytic Capacitor - A capacitor with two aluminum electrodes (the anode has the dielectric film) separated by layers of absorbent paper saturated with the operating electrolyte. The aluminum-oxide film or dielectric is reparable in the presence of an operating electrolyte.

Ambient Temperature - The temperature of the air or liquid surrounding any electrical part or device. Usually refers to the effect of such temperature in aiding or retarding removal of heat by radiation and convection from the part or device in question.

Anode - Positive electrode of a capacitor.

Bias Voltage - A voltage, usually DC, used to set the operating point of a circuit above or below a reference voltage.

Blocking Capacitor - A capacitor which limits the flow of DC or low frequency AC without materially affecting the flow of high frequency AC.

Breakdown Voltage - The voltage between the electrodes (plates) of a capacitor at which electric breakdown occurs under the prescribed test conditions. Also called dielectric breakdown voltage.

Bypass Capacitor - A capacitor which provides a low impedance path around a circuit element.

Capacitance - Property of a capacitor which determines its ability to store electrical energy when a given voltage is applied, measured in farads, microfarads, or picofarads.

Capacitive Reactance (X_C) - The opposition offered to the flow of an alternating or pulsating current by a capacitor or any unit having capacitance. Measured in ohms.

Capacitance Tolerance (Accuracy) - This tolerance is stated as the maximum positive and negative deviations of the capacitance from a rated nominal value, at a standard test temperature, measured at a standard frequency, and at a negligibly low test voltage. It is usually expressed as a percentage of nominal capacitance.

Capacitor - An electrical device capable of storing electrical energy and releasing it at some predetermined rate at some predetermined time. The capacitor consists of two conducting surfaces (electrodes) separated by an insulating (dielectric) material. A capacitor stores electrical energy, blocks the flow of direct current, and permits the flow of alternating or pulsating current to a degree dependent on the capacitance and the frequency. The amount of energy stored is expressed as: $E = {}^1\!/_2\, CV^2$

>*Liquid-filled* - A capacitor in which a liquid impregnant occupies substantially all of the case volume not required by the capacitor element and its connections. Space may be allowed for the expansion of the liquid under temperature variations.

>*Liquid-impregnated* - A capacitor in which a liquid impregnant is dominantly contained within the foil and paper winding, but does not occupy substantially all of the case volume.

>*Temperature-compensating* - A capacitor whose capacitance varies with temperature in a known and predictable manner.

Capacitor Bank - A number of capacitors connected together in series, in parallel or in series-parallel.

Capacitor-Input Filter - A power supply filter in which a capacitor is connected directly across, or in parallel with, the rectifier output.

Cathode - Negative electrode of a capacitor.

Charge - The quantity of electrical charge stored in a capacitor.

Coulomb - Unit of electric charge equal to the quantity of electricity transferred by a current of one ampere in one second.
1 coul. = 6.28×10^{18} electrons.

Coupling Capacitor - A capacitor used to transfer signals of a specified frequency from one circuit to another circuit.

Curie Point - In ferro electric dielectrics, the temperature(s) at which the dielectric constant reaches peak values. At the curie point temperature(s) the crystal form is changing from cubic to tetragonal. Manufacturers use modifiers in their specific ceramic formulations to shift curie points such that the rapid change in capacitance value which normally occurs around the curie point will have no effect on the specified characteristics of the capacitor over its specified temperature range.

CV Product - The capacitance of a capacitor multiplied by its rated voltage. A useful rating to compare capacitor technologies.

Cycle - The change of an alternating wave from zero to a negative peak to zero to a positive peak and back to zero. The number of cycles per second (cps or Hertz) is called the frequency.

DC Leakage Current - Stray direct current of relatively small value which flows through or across the surface of solid or liquid insulation when a voltage is impressed across the insulation.

Dielectric - The insulating (nonconducting) material (e.g., air, paper, mica, oil, etc.,) between the two electrodes (plates) of a capacitor.

Dielectric Absorption - Property of an imperfect dielectric whereby all electric charges within the body of the material caused by an electric field are not returned to the field. A measure of the reluctance of a capacitor dielectric to discharge completely. The charge remaining after a fully charged capacitor is momentarily discharged is expressed as a percentage of the original charge. Dielectric absorption is affected by charge/discharge time, voltage, and temperature.

Dielectric Breakdown Voltage - The voltage between the electrodes (plates) of a capacitor at which electric breakdown occurs under prescribed test conditions. Also called breakdown voltage.

Dielectric Constant (K) (Permittivity) - Property of a dielectric material that determines how much electrostatic energy can be stored per unit volume when unit voltage is applied. It is the ratio of the capacitance of a capacitor filled with a given dielectric to that of the same capacitor having a vacuum dielectric.

Dielectric Strength (Breakdown Voltage) - Maximum voltage that a dielectric material can withstand without rupturing or a conductive path is formed through (or around) it. The value obtained for the dielectric strength will depend on the thickness of the material and on the method and conditions of test. The dielectric strength is the ratio of the breakdown voltage to the thickness of the dielectric. In encapsulated units, it is also influenced by the coating material used. Unencapsulated units of voltages around 1KV DC and higher are generally not recommended, unless special design criteria is used to prevent arc-over between the electrodes through the air surrounding the part.

Dissipation Factor (DF, Tangent Delta) - The tangent of the dielectric loss angle. The ratio of the effective series resistance of a capacitor to its reactance at a specified frequency. Measured in percent.

Electrolyte - Current-conducting solution (liquid, gel, or solid) between two electrodes or plates of a capacitor at least one of which is covered by a dielectric film.

Electrolytic Capacitor - A capacitor consisting of two conducting electrodes whose anode has a metal oxide film on it. The film acts as the dielectric or insulating medium.

Electronic Industry Association (EIA) - An industry sponsored body, comprised of Manufacturers, Users and Equipment Manufacturers, which sets standards for electronic components and sets standards for packaging of components, measurement methods of components, handling methods for components, etc.

Equivalent Series Resistance (ESR) - For purposes of calculation, all internal AC series resistances of a capacitor (i.e., lead resistance, termination losses or dissipation in the dielectric material) treated as one single resistor. The square root of the difference between the impedance squared and the reactance squared.

Etching - An electro-chemical process that increases the surface area of aluminum or tantalum foil.

Farad - The basic unit of measure in capacitors. A capacitor charged to one volt with a charge of one coulomb (one ampere flowing for one second) has a capacitance of one farad. Capacitors are generally rated at portions of a Farad (microfarads or picofarads).
1 Farad (F) = 1,000,000 microfarads (μF).

Flashpoint of Impregnant - The temperature to which the impregnant (liquid or solid) must be heated in order to give off sufficient vapor to form a flammable mixture.

Impedance (Z) - Total opposition offered to the flow of an alternating or pulsating current, measured in ohms. Impedance is the vector sum of the resistance and the capacitive reactance, i.e., the complex ratio of voltage to current.

Impregnant - A substance, usually liquid, used to saturate paper dielectric and to replace the air between its fibers. Impregnation also increases the dielectric strength and the dielectric constant of the assembled capacitor.

Inductive Reactance (X_L) - Opposition offered to the flow of alternating current by coils, leads, solder connections or other inductors.

Insulating Sleeve - Tube or tape of insulating material placed around metal-enclosed capacitors to electrically insulate the case from the other components, wiring mounting rings, and the chassis of the equipment.

Insulation Resistance (IR) - Direct current resistance between two conductors that are separated by an insulating material. The ratio of the DC voltage applied to the terminals of a capacitor and the resulting leakage current through the dielectric and over its surface after the initial charging current has ceased. Specifications usually call for a certain minimum value (several thousand MegOhms) determined by the application of a specific voltage. It includes both the volume and surface resistance. Note: Capacitors are commonly subjected to two insulation resistance tests. One test determines the insulation resistance from terminal to terminal while the other test determines the insulation resistance from one or more terminals to the exterior case or insulating sleeve. Industry capacitor specifications typically performed after some two minutes charging time. For many types of capacitors it is expressed as the product of insulation resistance (IR) and capacitance values (MegOhms-microfarad).

Joule - A unit of energy or work. One joule is equal to one watt-second. Energy stored in a capacitor is equal to $CV^2/2$ joules or watt-seconds, where C is capacitance in farads and V is voltage at the terminals in volts.

KiloHertz (KHz) - Unit of frequency, 1,000 (10^3) cycles per second.

Leakage Current - Stray direct current of relatively small value which flows through capacitor when voltage is impressed across it.

Life Test - An accelerated test, designed to measure the ability of a capacitor to withstand its rated operating conditions for a lengthy, useful life. The test conditions depend upon the rating of the capacitor.

Mean Time Between Failures (MTBF) - MTBF defines the frequency of failure occurring in a large number of components in a system or a group of systems. MTBF cannot be used to predict failure of a single capacitor. Standard statistical procedures can be used to calculate the MTBF for systems with components with varying failure rates. The MTBF becomes an important design tool for determining component and systems reliability requirements. A specific MTBF figure can be calculated from the capacitor failure rate as follows: MTBF = 10^5/FR, where FR=failure rate in %/1000 hours, and MTBF = mean time between failures in unit-hours.

MegaHertz (MHz) - Unit of frequency, 1,000,000 (10^6) cycles per second.

Multilayer Capacitor - A ceramic capacitor, made up of several alternately stacked electrodes, separated by ceramic dielectric layer, fired into a single homogenous package.

Parts Per Million (PPM) - PPM defines error rates for quality data. This is calculated from electrical performance data including the minor or catastrophic variations. Today, not all suppliers are using a standard method of PPM calculation. Consequently, when comparing reported PPM levels, it is essential that the method of calculation be understood. For example, calculations that include only catastrophic failures may produce very low reported PPM levels.

Phase - The angular relationship between current and voltage in AC circuits. The fraction of the period which has elapsed in a periodic function or wave measured from some fixed origin. If the time for one period is represented as 360° along a time axis, the phase position is called the phase angle.

Polarized Capacitor - An electrolytic capacitor in which the dielectric film is formed on only one metal electrode. The impedance to the flow of current is then greater in one direction than in the other. Reversed polarity can damage the part if excessive current flow occurs.

Power Factor (PF) - The ratio of resistance to impedance, measured in percent.

Quality Factor (Q) - The ratio of capacitive reactance to its equivalent series resistance i.e., $Q = 1/DF$ or X_C/R.

Radio Interference - Undesired conducted or radiated electrical disturbances, including transients, which may interfere with the operation of electrical or electronic communications equipment or other electronic equipment.

Rated Capacitance - The value which is indicated on the capacitor. The actual capacitance value may deviate from this value within the tolerance limits for that capacitor.

Rated Working Voltage - The voltage which is indicated on the capacitor and which may be applied continuously to the terminals of the capacitor at temperatures within the applicable temperature category. Operation below the rated voltage (voltage derating) has a positive effect on the operational life.

Reactance (X) - Opposition to the flow of alternating current.

Reliability - The probability that a device will perform adequately for the length of time intended and in the operating environment encountered.

Resonant Frequency (f_o) - The frequency at which the total inductive and capacitive reactance of a capacitor (or of the components in a circuit) are equally low. This results in the component's impedance being equivalent to a pure resistor.

Reverse Leakage Current - A nondestructive current flowing through a polarized capacitor subjected to a voltage of polarity opposite to that normally specified.

Ripple Voltage (or Current) - The AC component of a uni-directional voltage or current. The AC component is usually small in comparison with the DC component. Ripple current is the total amount of alternating and direct current that may be applied to an electrolytic capacitor under stated conditions. In general, the higher the ripple current, the shorter the operating life of the capacitor. Application of significantly higher than rated ripple currents can shorten capacitor life dramatically and may even cause catastrophic failure, i.e., venting.

Stability - The ability of a component or device to maintain its initial operating characteristics after being subjected to changes in temperature, environment, current, and time. It is usually expressed in either percent or parts per million for a given period of time.

Surge Voltage (or Current) - The transient variation in the voltage or current at a point in the circuit; a voltage or current of large magnitude and short duration caused by a discontinuity in the circuit. The surge voltage is the maximum safe voltage to which a capacitor should be subjected under any combination of circumstances for a short period of time.

Temperature Coefficient (TC) - Change in capacitance of a capacitor per degree change in temperature. It may be positive, negative, or zero and is usually expressed in parts per million per degree Celsius (ppm/°C).

Time Constant - In a capacitor-resistor circuit, the number of seconds required for the capacitor to reach 63.2% of its full charge after a voltage is applied. The time constant of a capacitor with a capacitance (C) in farads in series with a resistance (R) in ohms is equal to R x C seconds.

Tolerance - The percentage of maximum deviation from the nominal capacitance value at a standard temperature, voltage, and frequency.

Voltage - Electrical pressure, i.e., the force which causes current to flow through an electrical conductor. The difference of potential between any two conductors. Expressed in volts.

Watt-Second - A unit of measure of electrical energy; the work done by one watt acting for one second. One watt-second is equal to one joule.

Working Voltage - The maximum voltage to be applied to a capacitor for continuous duty operation at maximum rated temperature.

Bibliography

Many thanks to the following who gave permission to use their information for reference, or for incorporation into this book:

AVX Corporation, Myrtle Beach, SC 29577: *Multilayer Ceramic Leaded Capacitors Data Book* (3902.5MCLC-C), *Glass Dielectric Capacitors Data Book* (7892.5MGDC-R).

Meedijk, V., *"Selecting the Best Resistor/Capacitor."* Radio Electronics, Gernsback Publications, Inc., Farmingdale, NY 11735, March, 1985, pp. 64-65, 109.

Mepco/Centralab, A North American Phillips Company, Riviera Beach, FL 33404: *1988-89 Resistor/Capacitor Data Book* (Form 87-002).

Sprague Electric Company, Longwood, FL 32750: *1990 Aluminum Capacitors Catalog* (AL-100), *1989 Tantalum Capacitors Catalog* (TA-100).

Union Carbide Corporation, Electronics Division (Kemet), Greenville, SC 29606: Engineering Bulletin (F-2856B 7/80), *What Is A Capacitor?*, *General Catalog* (F-2644T 6/85).

Additional Reference:

United States Military Standard: MIL-STD-198E NOTICE 2, *Capacitors, Selection and Use of*, September 16, 1988.

Appendix A

*Capacitor Selection Guidelines**

Ceramic

Values: 1 pF to 2.2 µF
Tolerance: 10% or 20%
Voltage rating: 3.3 volts to 6 Kilovolts DC
Dissipation factor: to 5%
Temperature coefficient: to 200,000 PPM/°C
Tolerance (For NP0's): 0.25% to 10%
Temperature coefficient: 0 ±30 and 0 ±60 PPM/°C
Notes: General purpose high insulation-resistance devices used for the transient decoupling of IC's and compensation of reactive changes caused by temperature variations. Applications include filtering, bypass, and noncritical coupling in high frequency circuits. Frequency sensitive (capacitance will vary with frequency) so characteristics should be measured at intended operating frequency. Should be mounted next to components being compensated, and shielded from sources of heat. Due to low voltage failure problems, should not be operated significantly under rated voltage under humid conditions. In circuit design, considerations should be given to changes in the dielectric constant caused be temperature, electric field intensity, and shelf aging.

Ceramic Chips

Values: 10 pF to 0.18 µF
Tolerance: 5% to 20%
Temperature range: -55°C to +125°C
Insulation resistance: greater than 100,000 Megohms

---

* Reprinted with permission from **Radio-Electronics** Magazine, March 1985 issue.
© Copyright Gernsback Publications, Inc., 1985.
These guidelines are to be used only as a reference for general capacitor characteristics. Refer to Manufacturer's data books for specific capacitor ratings and specifications.

Paper/Plastic Dielectric

Many dielectric and case configurations are available. Each type has its own characteristics. For example, metallized paper units have low insulation resistance and are prone to dielectric breakdown failures. Plastic types have superior moisture characteristics than paper units. Polycarbonate and Mylar types are used in applications that require minimum capacitance change with temperature, such as tuned or timing circuits.

Metallized polycarbonate and polycarbonate film

Values: up to 50 µF
Voltage rating: to 1000 WVDC
Dissipation factor: 0.5% (at 25°C and 120 Hz.)
Temperature range: -55°C to +125°C
Derating factors: 50% voltage; 80% of rated temperature
Notes: DC blocking, filter, bypass, coupling, and transient suppression applications. Close tolerance, high frequency capability (40 - 400 KHz) and high insulation resistance. Not suitable for sample/hold circuits, fast settling amplifiers, or filters due to dielectric absorption characteristics. Small size, medium stability, and long life expectancy under load.

Metallized polyester/polyester foil

Values: 0.001 to 100 µF
Voltage rating: up to 1500 WVDC
Dissipation factor: 1% (at 25°C and 120 Hz)
Temperature range: -55°C to +125°C (with 50% derating above 85°C)
Notes: See polycarbonate for typical applications. Moisture resistant and high insulation resistance. Small size, medium stability and very good load life. Capacitance will however vary widely with temperature. Foil units are generally lower cost than metallized types. Polyester film is commonly known as Mylar, which is a DuPont trademark.

Polystyrene foil

Values: to 10 µF
Voltage rating: up to 1000 WVDC
Dissipation factor: 0.03% (at 25°C and 120 Hz)
Temperature range: -40°C to +85°C without derating
Notes: Used in timing, integrating, and tuned circuits. High insulation resistance, and small capacitance change with temperature. Has excellent dielectric absorption characteristics. Large size with excellent stability and very good load life.

Paper/metallized paper/paper foil

Values: to 100 µF
Voltage rating: to 5000 WVDC
Temperature range: -30°C to +100°C (derated by 30% over 75°C)
Temperature coefficient: greater than 4,500 PPM/°C
Notes: General purpose. Medium stability and very good load life. Large size; low cost. Metallized paper has paper coated with thin layer of zinc or aluminum and are smaller than metal foil units. They are, however, prone to dielectric breakdown of insulation resistance and have poor surge handling capability. Paper foil units used in high voltage/high current applications. Their dissipation factor varies with temperature. Maximum temperature is +125°C.

Polypropylene foil/metallized polypropylene

Values: to 10 µF
Voltage rating: to 400 volts DC and 270 volts AC (foil units: 200 to 1600 volts DC and 300 to 440 volts AC)
Temperature range: -55°C to +105°C
Notes: Foil units are used in tuned circuits, integrating circuits, timing circuits, and CRT deflection circuits. Metallized units are used in DC blocking circuits. Good high frequency capability, high insulation resistance, close tolerance, high stability, and excellent dielectric absorption characteristics.

Less common types

Polysulfone
Similar to polycarbonate and polypropylene capacitors. Small size, high temperature range (to 150°C), suitable for high-frequency applications, and high insulation resistance. Excellent in high current and military applications. Not for sample/hold, fast settling amplifiers, or filters due to dielectric absorption characteristics. Poor history of availability.

Polyvinylidene fluoride
Considered experimental. Has high dielectric constant (about four to twelve times that of polyester devices), which results in a very small sized capacitor. These units suffer from significant capacitance change with temperature, particularly at low temperatures.

Polyethylene terephthalate
For applications that require high reliability and high insulation resistance at high temperatures.

Metallized paper polyester/paper polyester foil
The foil unit has a slightly better dissipation factor than the metallized type. Operating temperature of -55°C to +125°C with voltage ratings of 240 to 600 volts (DC) available.

Paper polypropylene
Available in voltage ratings of 400 to 800 volts (AC). Operating temperature from -40°C to +80°C.

Teflon/Kapton
Has a temperature range of -55°C to +250°C with a temperature coefficient of 0.009%/°C. Teflon's extremely low dielectric absorption makes it good for critical sample and hold circuitry. These capacitors used in specialized applications such as oil well drilling equipment. These capacitors are large in size since the dielectric is not available in thin gauges.

Parylene
Manufactured by Union Carbide, these capacitors are equivalent to polystyrene types in performance but are rated to +125°C, versus +85°C for polystyrene.

Aluminum Electrolytic

Values: 0.68 to 220,000 µF
Tolerance: -10% to +75%
Voltage rating: up to 350 volts
Temperature range: -55°C to +85°C (if derated, to +125°C)
Dissipation factor: varies with temperature
Temperature coefficient: varies with temperature
Notes: Used in filter, coupling, and bypass applications where large capacitance values are required and capacitances above nominal can be tolerated. Sum of the applied AC peak and DC voltages should never exceed the rated DC voltage. Aluminum electrolytics are larger than tantalum electrolytics but less expensive. Loss of capacitance, to as little as 10% of rated value, will occur as the aluminum oxide electrode electrochemically combines with the electrolyte. Oxide film deterioration also requires capacitors to be "re-formed" after storage to prevent dielectric failure. That involves application of rated voltage for a period of 30 minutes, or more, to restore initial leakage current value. Over time, dissipation factor can rise by as much as 50%. Four terminal devices are available (two leads for each connection) that offer low ESR and inductance at high frequencies. These units were designed for use in switching power supplies.

Tantalum Electrolytic

Solid type
Values: 0.001 to 1000 µF
Temperature range: -55°C to +85°C (if derated, to +125°C)
Voltage rating: 6 to 120 volts DC
Tolerance: 5% to 20%
Leakage current: varies with temperature
Derating factor: 50% voltage
Notes: Used in low-voltage DC applications such as bypass, coupling, and blocking. Not for use in RC timing circuits, triggering systems, or phase shift networks due to dielectric absorption characteristics. Also not recommended for applications subject to voltage spikes or surges. High capacitance in a small volume with excellent shelf life. Solid types not temperature sensitive and have lowest capacitance-temperature characteristic of any electrolytic unit. Dielectric absorption and high leakage currents make them unsuitable for timing circuits. Except for nonpolarized units, these devices should never be exposed to DC or peak AC voltages in excess of 2% of their rated DC voltage. To prevent failures due to leakage or shorting when series connecting for higher voltages, parallel each unit with a shunt resistor.

Chip types
Values: 0.068 to 100 µF
Tolerance: 5% to 20%
Voltage rating: 3 to 50 volts DC
Temperature range: -55°C to +125°C
Leakage current: varies with temperature

Nonsolid types
Values: 0.5 to 1200 µF
Tolerance: -15% to +30%, and 20%
Voltage rating: to 350 WVDC
Temperature range: -55°C to +85°C (if derated, to +125°C)
Leakage current: varies with temperature
Notes: Polarized foil units are used for bypassing or filtering out low-frequency pulsating DC. Allowance must be made for leakage current. Not suitable for timing or precision circuits due to wide tolerances. Large values available. Etched foil has 10 times the capacitance per unit volume as plain foil types. Peak AC and applied DC voltages should not exceed rated maximums. Usable to 200 KHz. Nonpolarized foil are used in tuned low- frequency circuits, phasing low-voltage AC motors, and in servo systems. Sintered slug units are used in low-voltage power supply filtering and in DC applications. Can not withstand any reverse voltage. Leakage current lowest of all tantalum types; no appreciable leakage below 85°C. Usable to frequencies of 1 MHz.

Glass

Values: 0.5 to 10,000 pF
Tolerance: to 5%
Voltage rating: 100 to 500 volts DC
Temperature range: -55°C to +125°C
Temperature coefficient: 0 to 140 PPM/°C
Notes: High insulation resistance, low dielectric absorption and fixed temperature coefficient. Has much higher Q than mica devices. Performs very well at high frequencies up to 500 MHz and can operate in range of 100 KHz to 1 GHz. Capable of withstanding severe environmental conditions but they are susceptible to mild mechanical shocks and should be mounted accordingly.

Mica

Values: 1 pF to 0.1 µF
Voltage ratings: 100 to 2500 volts DC
Temperature range: -55°C to +150°C
Temperature coefficient: -20 to +100 and 0 to +70 PPM/°C
Derating factor: 60% voltage (dipped case) and 40% voltage (molded case)

Mica Chips

Values: 1 to 10,000 pF
Voltage rating: to 500 volts
Notes: Used in timings, oscillator, tuned circuits, and where precise high frequency filtering is required. Capacitance and impedance limits are very stable and capacitors perform very well at frequencies of 10 KHz to 500 MHz. Devices using silver in their construction are very susceptible to silver ion migration resulting in short circuits. Failures can occur in a few hours if capacitors are exposed to DC voltage stresses, humidity, and high temperature.

Trimmer Capacitors

Values: Range from 0.25 to 1 pF and 1 to 120 pF
Glass/Quartz: Low loss, high Q, and high stability for high tuning sensitivity applications. Frequency range up to 300 MHz.
Sapphire: High level of performance between 1 and 5 GHz.
Plastic: High grade units can be operated up to 2 GHz.
Ceramic: Smallest sized single turn units with maximum capacitance under 100 pF. Capacitance changes with temperature.
Air: High level of performance through UHF band, from 300 MHz to 1 GHz.
Mica: Has wide capacitance range and relatively high current handling capability.
Vacuum/Gas: Used for high voltage applications. Values from 5 to 3000 pF, with voltage ratings from 2 to 30 Kilovolts (DC).

Appendix B

Equations and Symbol Definitions

Basic Capacitor Formulas

1. **Ohm's Law**

$$E = IR$$
$$P = EI$$

2. **Capacitance (farads)**

$$C = (8.85 \times 10^{-12}) K \frac{A}{D}$$

Where: K = Dielectric Constant
A = Area of Dielectric (Square Meters)
D = Distance Between Electrodes (Meters)

3. **Energy stored in capacitors (Joules, Watt-seconds)**

$$E = \tfrac{1}{2} C V^2$$

4. **Linear charge of a capacitor (Amperes)**

$$I = C \frac{dV}{dt}$$

5. **Total Impedance of a capacitor (Ohms)**

$$Z = \sqrt{R_s^2 + (X_L - X_C)^2}$$

6. **Capacitive Reactance (Ohms)**

$$X_C = \frac{1}{2\pi f C}$$

7. **Inductive Reactance (Ohms)**

$$X_L = 2\pi f L$$

8. **Phase Angles**

   Ideal Capacitors: Current leads voltage 90°
   Ideal Inductors: Current lags voltage 90°
   Ideal Resistors: Current in phase with voltage

9. **Dissipation Factor (%)**

$$D.F. = \tan\delta \text{ (loss angle)} = \frac{E.S.R.}{X_C} = (2\pi f C)(E.S.R.)$$

10. **Power Factor (%)**

$$P.F. = \sin\delta \text{ (loss angle)} = \cos\phi \text{ (phase angle)}$$

$$P.F. = D.F. \text{ (when less than 10%)}$$

11. **Quality Factor (dimensionless)**

$$Q = \cotan\delta \text{ (loss angle)} = \frac{1}{D.F.}$$

12. **Equivalent Series Resistance (ESR) (Ohms)**

$$E.S.R. = (D.F.)(X_C)$$

13. **Power Loss (Watts)**

$$\text{Power Loss} = (2\pi f C V^2)(D.F.)$$

14. **KVA (Kilowatts)**

$$KVA = 2\pi f C V^2 \times 10^{-3}$$

15. **Temperature Characteristic (ppm/°C)**

$$\text{T.C.} = \frac{C_t - C_{25}}{C_{25}(T_t - 25)}$$

16. **Capacitance Drift (%)**

$$\text{C.D.} = \frac{C_1 - C_2}{C_1} \times 100$$

17. **Reliability of Ceramic Capacitors**

$$\frac{L_o}{L_t} = \left(\frac{V_t}{V_o}\right)^X \times \left(\frac{T_t}{T_o}\right)^Y$$

Historically for ceramic capacitors exponent X has been considered as 3. The exponent Y for temperature effects typically tends to run about 8.

18. **Capacitors in Series (current the same)**

$$\text{Any Number:} \quad \frac{1}{C_T} = \frac{1}{C_1} + \frac{1}{C_2} \cdots \frac{1}{C_N}$$

$$\text{Two:} \quad C_T = \frac{C_1 C_2}{C_1 + C_2}$$

19. **Capacitors in Parallel (voltage the same)**

$$C_T = C_1 + C_2 \cdots + C_N$$

20. **Aging Rate**

$$\text{A.R.} = \%\Delta C \,/\, \text{decade of time}$$

21. **Decibels**

$$db = 20 \log \frac{V_1}{V_2}$$

Appendix B

Metric Prefixes

Pico	$\times 10^{-12}$	Tera	$\times 10^{12}$
Nano	$\times 10^{-9}$	Giga	$\times 10^{9}$
Micro	$\times 10^{-6}$	Mega	$\times 10^{6}$
Milli	$\times 10^{-3}$	Kilo	$\times 10^{3}$
Deci	$\times 10^{-1}$	Deca	$\times 10^{1}$

Symbols

K = Dielectric Constant

T_D = Dielectric thickness

E = Voltage

R = Resistance

R_s = Series Resistance

L = Inductance

ϕ = Phase angle

L_t = Test life

V_o = Operating voltage

T_o = Operating temperature

A = Area

V = Voltage

I = Current

t = time

f = frequency

δ = Loss angle

L_o = Operating life

V_t = Test voltage

T_t = Test temperature

P = Power

X & Y = exponent effect of voltage and temperature

Index

A

AC
  definition, 107
AC ripple, 17
Aging
  definition, 107
Alternating current, 11
  sine wave, 11
Aluminum electroltyic capacitor, 57 - 78
Aluminum electrolytic capacitor
  bathtub curve, 70
  case to ambient thermal resistance, 68
  computer grade, 74
  construction, 58
  energy storage capabilities, 60
  multi-tabbed construction, 66
  nonpolar, 65
  production technology, 61
  reliability, 74
  ripple current capabilities, 67
  thermal efficiency of, 67
  thermal resistance equation, 68
Aluminum foil, 49
Ambient temperature
  definition, 107
Anode, 61, 80 - 81
  definition, 107

B

Barium titanate, 5, 41
Bias voltage
  definition, 107
Breakdown Voltage
  definition, 108, 110

C

Capacitance
  definition, 108
  equation, 4
  equation for multiple layers, 31
  rated definition, 114
  stray, 24
  temperature compensation, 24
  tolerance definition, 115
  units, 4
Capacitance tolerance
  definition, 108
Capacitive reactance, 13
  definition, 108
  effects of frequency, 13
  equation, 13, 126
Capacitor
  acid filled, 26
  aging, 36, 41
  aluminum electrolytic, 57 - 78
  aluminum electrolytic definition, 107
  balancing resistors, 25
  blocking definition, 107
  bypass, 21
  bypass definition, 108
  ceramic chip, 43
  ceramic disc, 29, 39
  ceramic trimmer, 43
  Class 1, 33
  Class 2, 33
  coupling definition, 109
  DC filter, 8
  definition, 108
  design criteria, 24, 39
  dipped disc, 30
  effects on peak voltage, 24
  electrolytic definition, 111
  energy storage equation, 57
  energy storage in, 21, 57
  improved DC filter, 10
  infant failure, 70
  input filter definition, 109
  insulating sleeve, 76
  internal charge, 3
  life, 22
  liquid filled, 26
  liquid filled definition, 108
  liquid impregnated definition, 108
  mechanical effects, 22, 36
  metallized film, 50

monolithic, 30
multilayer definition, 113
oil filled, 26
parallel plate, 1, 47
piezoelectric effect, 36
plastic film, 47 - 56
polarized definition, 113
power equation, 18
power handling, 16
practical, 15
radiation field, 28
schematic symbol, 1
selection considerations, 21
selection guidlines, 119
self-healing, 25, 51, 55
surface temperature rise, 25
temperature compensating, 32, 38
temperature compensating chart, 34
temperature compensating definition, 108
trimmer, 26
water flow analogy, 2
Capacitor bank
definition, 109
Cathode, 65, 80
definition, 109
Ceramic dielectric
volumetric efficiency of, 32
Ceramic disc
construction of, 30
Charge
definition, 109
Coulomb
definition, 109
Curie point, 36, 41
definition, 109
CV product
definition, 109
Cycle
definition, 109

D

DC leakage current
definition, 109
De-aging, 36
Dielectric
definition, 109
effects of moisture, 23
effects of pressure, 23
effects on polarized, 22
introduction, 1
strength, 22
temperature effects, 21
Dielectric absorption, 93
definition, 110
Dielectric breakdown voltage
definition, 110

Dielectric constant, 4
definition, 110
table, 6
Dielectric strength
definition, 110
Dissipation factor, 16
definition, 110
equation, 16

E

EIA, 33
definition, 111
RS-198 specification, 33
EIA codes, 35
C0G, 32
chart, 35
NP0, 32
X7R, 32
Z5U, 32
Electrolyte, 63
definition, 110
Equivalent series resistance (ESR), 16, 63
ESR
definition, 111
Etching, 59
definition, 111

F

Farad
definition, 111
Faraday, Michael, 4
Filters
LC filter, 10
Flashpoint of impregnant
definition, 111
Frequency
cycles per second, 13
Hertz, 13

G

Glass capacitor, 97 - 102
advantages of, 98
aging rate, 100
applications, 101
construction, 97
dielectric absorbtion in, 98

I

Impedance, 14
definition, 111
determining, 14
Impregnant
definition, 111
Inductance
units, 13

Inductive reactance, 13
    definition, 111
    effects of frequency, 13
    equation, 13, 126
Infant failure, 70
Insulating sleeve
    definition, 112
Insulation resistance, 21, 39, 53
    definition, 112

J

Joule
    definition, 112

K

K, 4
KiloHertz
    definition, 112

L

Leakage current
    definition, 112
Life test
    definition, 112

M

Manganese dioxide, 81
Mean time between failures (MTBF)
    definition, 112
MegaHertz
    definition, 113
Metallized film, 50
Mica capacitor, 103 - 106
    button style, 103
    characteristics, 105
    construction, 103
    silver ion migration, 105

O

Ohm's law, 17
    equation, 17, 125
    for AC circuits, 18

P

Parts per million (PPM)
    definition, 113
Passive components, 2
Permittivity, 4
    of vacuum, 4
Phase
    definition, 113

Piezoelectric, 36
Plastic film, 47
    construction, 47
Plastic film capacitor, 47 - 56
Polycarbonate, 49
Polyester, 53
Polyethylene terephthalate, 49
Power
    equation, 17, 125
Power factor
    definition, 113

Q

Quality factor (Q)
    definition, 113

R

Radio interference
    definition, 113
Rated working voltage
    definition, 114
RC timing circuit, 6
    example, 7
Reactance
    definition, 114
Reliability, 37
    definition, 114
Resonant frequency
    definition, 114
Reverse leakage current
    definition, 114
Ripple voltage
    definition, 114
    equation, 18
rms voltage, 19
    relationship of peak and rms, 19

S

Scintillation, 63
Self-healing
    in plastic film capacitors, 51
Self-resonant point, 15
Silver ion migration, 40
Sintering, 83
Spacer, 65
Stability
    definition, 114
Surge voltage, 73
    definition, 115

T

Tantalum, 32
Tantalum capacitor, 79 - 96
   chip, 95
   encapsulation of, 85
   foil, 79 - 80
   foil construction, 80
   healing, 85
   nonpolarized applications, 89, 95
   oxide forming chemical equation, 84
   solid, 81
   solid construction, 82
   wet, 81
   wet-slug, 79
   wet-slug construction, 81
Tantalum pentoxide, 84
Temperature coefficient (TC)
   definition, 115
Time constant
   definition, 115
Titanium dioxide, 42

V

Voltage
   average, 19
   bias, 19
   definition, 115
   polarizing, 41

W

Watt-second
   definition, 115
Working voltage
   definition, 115

Notes

Made in the USA
Charleston, SC
19 September 2011